Emerging Technologies in Agriculture and Food Science

Edited by

Karim Ennouri

University of Sfax,
Sfax,
Tunisia

Emerging Technologies in Agriculture and Food Science

Editor: Karim Ennouri

ISBN (Online): 978-981-14-7000-4

ISBN (Print): 978-981-14-6998-5

ISBN (Paperback): 978-981-14-6999-2

need for a court order if at any point you breach any terms of this License Agreement. In no event will any delay or failure by Bentham Science Publishers in enforcing your compliance with this License Agreement constitute a waiver of any of its rights.

3. You acknowledge that you have read this License Agreement, and agree to be bound by its terms and conditions. To the extent that any other terms and conditions presented on any website of Bentham Science Publishers conflict with, or are inconsistent with, the terms and conditions set out in this License Agreement, you acknowledge that the terms and conditions set out in this License Agreement shall prevail.

Bentham Science Publishers Pte. Ltd.
80 Robinson Road #02-00
Singapore 068898
Singapore
Email: subscriptions@benthamscience.net

BENTHAM SCIENCE

CONTENTS

FOREWORD

I am delighted to write the foreword of the book titled *"Emerging Technologies in Agriculture and Food Science"* edited by Dr. Karim Ennouri and published by Bentham Science Publishers. I know Dr. Karim Ennouri for more than ten years, and I deeply believe in the research value of interpretive discussion in the biotechnology domain.

Biotechnology is considered as the modern green revolution, offering influential instruments for efficient advanced crop plants, in addition to other organisms through constantly growing technologies aimed at well-organized employment of biological systems to benefit humanity. Applied biotechnology presents an exceptional occasion to propagate scientific perception of a variety of dynamic phenomena and processes related to ecosystems. The exploitation of data sets and the improvement of original data processing algorithms assist in developing aptitudes to process all dimensions of plant observation data and employ these data in making management verdicts and decisions.

I hope and expect that this book will provide an effective learning experience and referenced resource on the topics of agro-biotechnology, bioactive elements, monitoring of vegetation dynamics and modeling, and biotechnological innovations of natural products.

Faiçal Brini
Biotechnology and Plant Improvement Laboratory
Centre of Biotechnology of Sfax
Tunisia

PREFACE

Nowadays, cultivators are increasingly arranging innovative, highly technical and scientific estimations with the aim of enhancing agricultural sustainability, effectiveness, and/or plant health. Innovative farming technologies incorporate biology with smart technology: Computers and devices exchange with one another autonomously in a structured farm management system. Throughout this structure, smart agriculture can be accomplished; cultivators decrease plantation inputs (pesticides and fertilizers) and increase yields via integrated pest management and/or biological control.

Moreover, the intensive use of pesticides creates imbalances in the microbial community, which may be unfavorable for the activity of the beneficial organisms and may also lead to the development of resistant pathogen strains, increasing environmental degradation. Owing to the limitations of chemical control measures, it seems appropriate to seek a more suitable control method. Biological control appears as the most promising strategy, being environmentally safe and cost-effective for controlling several phytopathogens. Therefore, the development of novel agents can be useful in the control of plant diseases. Recently, there has been a growing interest in researching the possible use of functional biomolecules that possess a selective action against these fungi without being toxic to the ecosystem for pest and disease control in agriculture. Natural biomolecules are increasingly becoming an effective and environmentally friendly tool for the control of phytopathogenic agents.

This book resumes present innovative techniques and methodologies to complement usual plant control and breeding attempts toward enhancing crop yield and production and consequently maintaining food security.

Karim Ennouri
University of Sfax
Sfax
Tunisia

LIST OF CONTRIBUTORS

Ahlem Chakchouk-Mtibaa Laboratoire de Microorganismes et de Biomolécules du Centre de Biotechnology de Sfax, Sfax, Tunisia

Ennio Ottaviani OnAIR Ltd, Genoa, Italy
Department of Mathematics, University of Genoa, Genoa, Italy

Emna Bouazizi Laboratoire Les Ressources Génétiques de l'Olivier: Caractérisation, Valorisation et Protection Phytosanitaire, Institut de l'Olivier, Université de Sfax, Tunisie

Enrico Barelli OnAIR Ltd, Genoa, Italy

Hajer Ben Hlima Unité de Biotechnologie des Algues, Biological Engineering Department, National School of Engineers of Sfax, University of Sfax, Sfax, Tunisia

Imen Sellem Laboratoire de Microorganismes et de Biomolécules du Centre de Biotechnology de Sfax, Sfax, Tunisia

Karim Ennouri Technopark of Sfax, Sfax, Tunisia
Olive Tree Institute, Sfax, Tunisia

Khaoula Elhadef Laboratoire de Microorganismes et de Biomolécules du Centre de Biotechnology de Sfax, Sfax, Tunisia

Lotfi Mellouli Laboratoire de Microorganismes et de Biomolécules du Centre de Biotechnology de Sfax, Sfax, Tunisia

Manel Cheffi Laboratoire Les Ressources Génétiques de l'Olivier: Caractérisation, Valorisation et Protection Phytosanitaire, Institut de l'Olivier, Université de Sfax, Tunisie

Mariam Fourati Laboratoire de Microorganismes et de Biomolécules du Centre de Biotechnology de Sfax, Sfax, Tunisia

Mohamed Ali Triki Olive Tree Institute, Sfax, Tunisie
Laboratoire Les Ressources Génétiques de l'Olivier: Caractérisation, Valorisation et Protection Phytosanitaire, Institut de l'Olivier, Université de Sfax, Sfax, Tunisia

Olfa Ben Braïek Laboratory of Transmissible Diseases and Biologically Active Substances (LR99ES27), Faculty of Pharmacy, University of Monastir, Tunisia

Paola Cremonesi Institute of Agricultural Biology and Biotechnology, Italian National Research Council (CNR IBBA), Lodi, Italy

Stefano Morandi Institute of Sciences of Food Production, Italian National Research Council (CNR ISPA), Milan, Italy

Slim Smaoui Laboratoire de Microorganismes et de Biomolécules du Centre de Biotechnology de Sfax, Sfax, Tunisia

Yaakoub Gharbi Laboratoire Les Ressources Génétiques de l'Olivier: Caractérisation, Valorisation et Protection Phytosanitaire, Institut de l'Olivier, Université de Sfax, Tunisie

CHAPTER 1

Active Compounds from Pomegranate Seed: New Source for Food Applications

Slim Smaoui[*, 1], **Mariam Fourati**[1], **Hajer Ben Hlima**[2], **Khaoula Elhadef**[1], **Olfa Ben Braïek**[3], **Ahlem Chakchouk-Mtibaa**[1], **Imen Sellem**[1] and **Lotfi Mellouli**[1]

[1] *Laboratoire de Microorganismes et de Biomolécules du Centre de Biotechnology de Sfax, Sfax, Tunisia*

[2] *Unité de Biotechnologie des Algues, Biological Engineering Department , National School of Engineers of Sfax, University of Sfax , Sfax, Tunisia*

[3] *Laboratory of Transmissible Diseases and Biologically Active Substances (LR99ES27) , Faculty of Pharmacy, University of Monastir, Tunisia*

Abstract: Pomegranate is an essential fruit bearing tree well cultivated in the world. Biological potential and nutritional value were very reputed in both of pomegranate fruit and its by-products, such as seeds. According to the presented information in literature, the use of pomegranate seed as a natural food preservative can be explained by its phytochemicals richness. Based on this phenolics content of pomegranate seed (PS) extracts, the current chapter will talk about its successful use as natural preservative agent in the development of healthier and shelf stable food products. This document speaking of the antioxidant and antimicrobial activities evaluation of PS and its principal active phenolic compounds identified and quantified by advances in the separation sciences and spectrometry, will perform a comprehensive review of the scientific literature. Furthermore, the impact of using PS on the food quality and agri-food products was also evaluated.

Keywords: Advanced analytical chemistry, Antioxidant and Antimicrobial activities, Biopreservation, Food and agri-food products, Pomegranate seed, Phenolic compounds.

INTRODUCTION

Because food and agri-food products preservation has become an international problem, the used chemicals are severely regulated [1, 2]. For that reason, in the previous decade, the exploitation of natural products with biological properties has been revived attention given its remarkable phytochemical and reliable

---

[*] **Corresponding author Dr. Slim Smaoui:** Laboratoire de Microorganismes et de Biomolécules du Centre de Biotechnology de Sfax, Sfax, Tunisia; E-mails: slim.smaoui@cbs.rnrt.tn; slim.smaoui@yahoo.fr

approaches for the preservation of food products. In this line, bioactive compounds of fruits and vegetables, including sources of flavonoids, phenolic acids and pigments, have been examined for enhancing human health and ensuring food security due to their biological potential [3 - 5].

Given its remarkable phytochemical content, pomegranate (*Punica granatum*) has captured increased interest [6 - 9]. Considered usually as waste, pomegranate seeds (PS) have been described as being abundant in polyphenols such as flavonoids (anthocyanins, catechins and other complex flavanoids) and hydrolyzable tannins (punicalin, pedunculagin, punicalagin, gallagic and ellagic acid esters of glucose) [10, 11]. Equally, PP's Phenolic compounds contain anthocyanins, gallotannins, ellagitannins, gallagylesters, hydroxyl benzoic acids, hydroxyl cinnamic acids and dihydro flavonol [12, 13]. Fig. (**1**) demonstrates the polyphenols structure found in pomegranate [13]. Phenolics of PS have been reported to display realistically elevated free radical scavenging activities and also powerful antimicrobial activity [14, 15]. Furthermore, their nutraceutical application, PS exposes important properties and techno-functional food applications, (*e.g.* antioxidant, antimicrobial, colorant and flavoring) [16 - 18]. PS, evenly, can proceed as notable natural additives for food and agri-food products and their quality development thus affording a well-founded alternative to synthetic antioxidants [18 - 20].

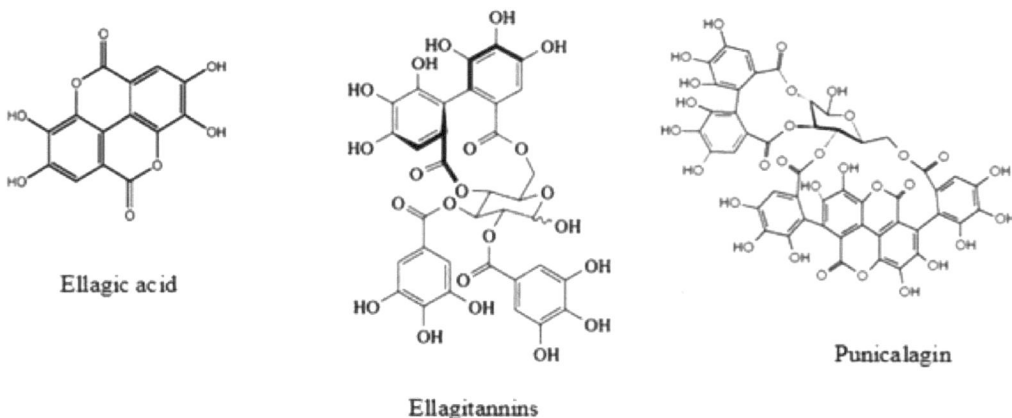

Ellagic acid

Ellagitannins

Punicalagin

Fig. (1). Structure of polyphenols present in pomegranate.

For that reason, the existing chapter presents a cumulative in-depth knowledge on the analytical techniques exploited in categorization of PS phenolic compounds, anti oxidant and antibacterial potentials and successful exploitation of PS in food and agri-food products.

BOTANICAL ASPECTS OF POMEGRANATE

Pomegranate probably originated from Saxifragales belonging to the order Myrtales [21]. It was in 1753 that the genus *Punica*, having tropical ancestors close to Lythraceae and Sonneratiaceae, was described for the first time by Linnaeus [22]. The evolution of Punica along the xero- and cryophilic lines of development, caused its Arogenesis. Punicaceae is a monogeneric family that includes a single genus *Punica* of two species, *Punica granatum* L. and P. *protopunica* Balf. f., (syn. Socotria protopunica) with the latter endemic to Socotra Island (Yemen) whereas *Punica nana*, another form of *P. granatum* is frequently considered as third species of *Punica* [23].

More than 1000 cultivars of *Punica granatum* are present [24], native from the Middle East, prolonged throughout the Mediterranean, eastward to China and India, and on to the American Southwest, California and Mexico in the New World. The fruit itself donates rise to three parts: the seeds, about 3% of the fruit weight, and themselves possessing around 20% oil, the juice, near 30% of the fruit mass, and the peels (pericarp) who also contain the inner network of membranes. Other functional parts of the plant comprise the roots, bark, leaves, and flowers [25].

PS Phytochemical Content

From Tunisian pomegranate fruits, PS total phenolic content (TPC) was 326.7 ± 1.4 mg gallic acid equivalent/100 g fresh matters (FM) [26]. These authors confirmed that this value was in accord with preceding finding in Indian PS, which ranged from 230 to 510 mg gallic acid equivalent (GAE)/100 g FM [27].

According to Elfalleh *et al.* [28], Tunisian PS extracted with methanol had higher TPC values (11.84 GAE mg/g dry weight), flavonoids (6.79 mg rutin equivalents per dry weight (mg RE/g DW/g DW), anthocyanins (40.84 mg of cyanidin---glucoside equivalents per g DW (mg CGE/g DW) and hydrolysable tannins (29.57 mg tannic acid equivalent per g of DW (mg TAE/g DW). Gozlekci *et al.* [29] investigated the total phenolic of four Turkish PS.

PS contained 3.3% of the overall fruit phenolic content and TPC was ranged from 117.0 to 177.4 mg GAE/L. In fact, TPC of PS from cultivars "Asinar," "Lefan," "Katirbasi," and "Cekirdeksiz-IV" was 177.4 mg/L, 125.3 mg/L, 121.2 mg/L, and 117.0 mg/L, respectively [24]. Kalaycıoğlu and Erim [30] calculated the levels of bioactive compounds and antioxidant activities in juice, peel and seed of 3 genotypes of pomegranate cultured in Turkey. Results discovered that the peel extract had about 12.4-fold higher total flavonoid than that of juice extract, and seed extract had 13.4-fold more total flavonoid than that of juice extract. Pande and Akoh [31] reported that TPC in Georgian PS are 365 mg GAE/g FW. In the study of Derakhshan *et al.* [32], TPC of Iranian PS extracted from three varieties of pomegranate was ranged between 72.4 and 73 mg GAE/g. The TPC of Iranian PS extracts assorted with solvent at a ratio of 1:10 (w/v) were calculated [33]. In this study, solvents were water, methanol, acetone, ethyl acetate and hexane. The authors indicated that the effectiveness of the solvents for extraction of the phenolic compounds was in the following order: methanol > water > acetone > butanol > ethyl acetate > hexane, which was confirmed by Singh *et al.* [34] results. These authors also described that the water extract of PS (3%, w/w) had the highest TPC, succeed by methanol extract (2.6%) and ethyl acetate extract (2.1%). However, methanol extract (27.93 mg/L seed extract) of PS presented highest phenolics and hexane extract (0.29 mg/L seed extract) showed the minimum of phenolics.

TPC of 50% aqueous acetone like extraction solvent assorted from 1.29 to 2.17 mg GAE/g in four pomegranate China cultivars [35]. These authors demonstrated that the TFC attained 80% aqueous methanol between 0.37 and 0.58 mg catechin equivalents (CAE) in PS. Extracted by 80% methanol, China PS restricted the highest TFC value (0.58 mg CAE/g) among these four cultivars. Total proanthocyanidins considerably wide-ranging from 68 to182 µg cyanidin equivalents/g seeds in the four seed samples [35].

Thailand PS extracted with ethanol showed a higher yield (11.9%) than that extracted with acetone (10.0%) extract [36].

For TPC, acetone extracts (0.175 mg GAE/g extract) of PS showed higher TPC than ethanol (0.136 mg GAE/g extract) and water (0.084 mg GAE/g extract) extracts. Among three extraction solvents extract, acetone extracts of PS had higher ($P < 0.05$) TFC [36].

TPC of the fraction involving free and esterified phenolics (soluble) from PS extracted from pomegranate fruits grown in California-USA were resolved by Folin–Ciocalteu's method.

The TPC value of PS was 3.39 mg GAE/g of sample. According to He *et al.* [37], the TPC of PS extracted with 70% acetone was 24.28 mg catechin equivalents per gram (dw).

Malaysian ethanol (70%) PS extract was analyzed for its TPC. In fact, TPC in PS was 165 ± 49 mg GAE/L [38].

These variations in TPC might be due to the divergence between the extraction methods; in fact, phenolic contents of PS extracts are projected to robustly depend on extraction circumstances as well as the sort of solvent used [38, 39].

The phenolic compound solubilities are depending on the polarization of solvents. They are extracted greatly in polar solvents (*e.g.* water and methanol), however, and these compounds are not extracted with nonpolar solvents.

Table **1** summarizes some results found in relation to this topic in recent years.

Advanced Analytical Chemistry of Phenolic Compound from PS

By using high-performance liquid chromatography-diode array detection-electrospray ionization/mass spectrometry (HPLC-DAD-ESI-MS/MS), a total of 47 phenolic compounds were identified in the extracts of PS [40].

Thirteen phenolic acids were recognized by UV spectra, the MS and literature data. According to Ambigaipalan *et al.* [40], protocatechuic acid, vanillic acid, gallic acid, brevifolin carboxylic acid, p-hydroxybenzoic acid hexoside, cis- and trans-caffeic acid hexoside, derivative of caffeic acid hexoside, vanillic acid hexoside, ferulic acid hexoside, catechin, quercetin hexoside, cis- and trans-dihydrokaempferol-hexoside, ellagic acid, ellagic acid pentoside, ellagic acid deoxyhexose and ellagic acid hexoside, valoneic acid bilactone, digalloyl hexoside, and galloyl-HHDP hexoside were known for the first time in PS. On the other hand, Fischer *et al.* [48] confirmed that cyanidin– pentoside–hexoside, valoneic acid bilactone, brevifolin carboxylic acid, vanillic acid 4-glucoside and dihydrokaempferol-hexoside are identified for the first time in pomegranate fruits.

Table 1. Phenolic contents reported in PS from different regions.

Source	Solvent Extraction	Analyte	Method	Quantification	Detection	Reference
Thailand	- Water (W) - Ethanol (95%) (E) - Acetone (70%) (A)	TPC (mg GAE/g)	Folin–Ciocalteu	0.084 (W) 0.136 (E) 0.175 (A)	765 nm	[36]
		Total flavonoid content (TFC) (mg quercetin equivalent per g (mg QE/g)	Aluminium chloride colorimetric technique	0.027 (W) 0.042 (E) 0.069 (A)	415 nm	

(Table 1) cont.....

Source	Solvent Extraction	Analyte	Method	Quantification	Detection	Reference
Iran	Ethanol (80%)	Total TFC (mg GAE/g)	Folin–Ciocalteu	72.4-73.13	765 nm	[32]
		TFC (mg rutin/g)	Aluminium chloride colorimetric technique	7.55-38	415 nm	
		TFC (mg rutin/g)	Aluminium chloride colorimetric technique	3.4-22	440 nm	
Canada+Brazil	Defatted with Hexane	TPC (mg GAE/g)	Folin-Ciocalteu	0.62-1.39	765 nm	[40]
		Monomeric anthocyanin contents (mg/100g)	pH-differential method	3.7-7.04	520 nm and 700 nm	
China	Acetone (70%)	TPC (mg/g d.w)	Folin–Ciocalteu	6.17-12.44	765 nm	[41]
Egypt	Water	Total phenolic (GAE mg/g FW)	Folin–Ciocalteu	8.67-12.37	765 nm	[42]
		TFC (RE mg/g FW)	Colourimetric method	19.84-26.45	440 nm	
Malaysia	Ethanol (70%)	TPC (mg GAE/L)	Folin–Ciocalteu	165	760 nm	[38]
Iran	Water (W) Methanol(M) Acetone (A) Butanol (B) Ethyl acetate(EA) Hexane (H)	TPC (mg/L seed extract)	Folin–Ciocalteu	22.61 (W) 27.93 (M) 3.41 (A) 0.57 (B) 0.37 (EA) 0.29(H)	760 nm	[33]
Tunisia	- Untreated seeds (US) - Osmotic Dehydration Treatment (ODT) - Air-Drying Experiment 40-50-60°C (ADE)	TPC (mg/100 g)	Folin–Ciocalteu	326.68 (US) 184.39 (ODT) 134.58-151.56 (ADE)	410 nm	[26]
		Total anthocyanin (mg/100 g)	pH differential method	82.3 (US) 68.43 (ODT) 20.1-40.11 (ADE)	510 nm and 700 nm	

(Table 1) cont.....

Source	Solvent Extraction	Analyte	Method	Quantification	Detection	Reference
Iran	- Ethanol - Chloroform - Water	TFC Tannins	- Ferric chloride test - Alkaline reagent Test - Lead acetate solution Test - Gelatin Test	(- in all tests) (+ in Ethanol)	760 nm	[43]
Tunisia	- Water (W) - Methanol (M)	Total polyphenol (GAE mg/g dry weight)	Folin–Ciocalteu	7.94 (W) 11.84 (M)	765 nm	[28]
		TFC (RE mg/g dry weight)	colourimetric method	3.35 (W) 6.79 (M)	430 nm	
		Total anthocyanins (CGE mg/g dry weight)	pH differential method	19.62 (W) 40.84 (M)	510 nm and 700 nm	
		Hydrolysable tannin (TAE mg/g dry weight)		32.86 (W) 29.57 (M)	550nm	
China	-Acetone (50%) (A) - Methanol (80%) (M) Acetone/ Methanol/Water	TPC (mg GAE/g)	Folin–Ciocalteu	1.29-2.17 (A) 1.29-1.79 (M)	765 nm	[35]
		TFC (mg CAE/g)	colourimetric method	0.37-0.58 (A) 0.42-0.62 (M)	510 nm	
China	Water (W) Methanol (M) Ethanol (E) Acetone (A)	TPC (mg/100 g DW)	Leaching extraction (all solvents)	1116-2194.3	765 nm	[37]
			ultrasound-assisted extraction (all solvents)	312.9-566.4		
			Soxhlet extraction (M) (A)	10.92-1965.7		
			Subcritical water extraction (120 – 160 - 180 °C) (W)	992.3-2752		
Thailand	- Ethanol (70%) (E) - Water (W)	TPC (mg GAE/mg)	Folin–Ciocalteu	77.93 (E) 51.58 (W)	760 nm	[44]
		TFC (mg GAE/mg)	Colorimetric method	16.66 (E) 10.55 (W)	510 nm	

(Table 1) cont.....

Source	Solvent Extraction	Analyte	Method	Quantification	Detection	Reference
Turkey	Water (W) Methanol (M) Ethanol (E)	TPC (μg gallic acid equivalents/mg extract)	Folin–Ciocalteu	49.67-56.89 (W) 54.64-63.77 (M) 54.48-56.89 (E)	760 nm	[45]
		TFC (μg QE /mg)	colourimetric method	18.73-24.94 (W) 5.92-15.59 (M) 9.35-16.92 (E)	430 nm	
		Total tannin content (μg tannic acid eqivalent/mg extract)	AOAC	16.32-29.07 (W) 32.43-56.01 (M) 10.45-25.14 (E)	760 nm	
Turkey	Ethanol/ Water 50% (v/v)	TPC (mg GAE/L)	Folin-Ciocalteu	117-177.4	765 nm	[29]
USA	Water	TPC (%)	Folin–Ciocalteu	Dry seeds 6% Wet seeds 5%	760 nm	[46]
USA	Hexane	TPC (mg GAE/ 100 g FW)	Folin–Ciocalteu	84.9-91.1	765 nm	[31]

The HPLC-DAD-ESI-MS/MS quantification of phenolic acids from PS demonstrated that gallic acid (~1037 μg/100 g dry weight) was the major phenolic acid present in PS, which followed the order of insoluble-bound > esterified > free. A total phenolic acid obtained from PS was 1164 μg/100 g DW. Eight flavonoids, namely (+)-catechin, dihydroxygallocatechin, naringenin hexoside, quercitrin-3-O-rhamnoside, quercetin hexoside, kaempferol-3-O-glucoside, as well as cis- and trans-dihydrokaempferol-hexoside were absolutely recognized in PS. Catechin was identified only in the free phenolic fraction of PS with its characteristic molecular ion [M-H]⁻ at *m/z* 289. Naringenin hexoside was detected in the free phenolic fraction of PS as its deprotonated molecular ion at *m/z* 433 [40].

Thirteen hydrolysable tannins were identified in PS [40]. Ellagic acid (m/z 301) and MS² fragments *m/z* 185, 229, 257 and 283 were identified in all fractions of PS based on the retention time, UV spectra and MS data of the authentic standard. Ellagic acid-derivative II (*m/z* 425) and ellagic acid pentoside (*m/z* 433) were detected as isomers of ellagic acid derivative in PS, respectively. Isomers of hexahydroxydiphenoyl (HHDP) hexoside have already been known in PS [49 - 51].

Corilagin or galloyl-HHDP hexoside was also recognized for the first time in PS by Ambigaipalan *et al.* [40].

Ellagic acid was the most important hydrolysable tannin present in PS (~ 220 µg/100 g dry weight), which followed the order of free > insoluble-bound > esterified. Twelve anthocyanins were detected in PS based on UV spectra (520 nm), MS^2 data in positive mode, as well as data from the literature. Four anthocyanins, namely cyanidin-3-O-pentoside, pelargonidin-3-O-glucoside, cyanidin-3-O-glucoside and delphinidin-3-O-glucoside were identified in PS [47].

In the study of He *et al.* [37], phenolic compounds were extracted and isolated from PS residue (PSR). TPC and proanthocyanidin (PC) contents of the extracts were concluded as 2427.90 and 505.63 mg catechin equivalent of 100 g/DW, respectively. Seventeen compounds in PSR extracts were discovered with antioxidant capacity and HPLC–ESI–MS was used to identify them. The main identified phenolics in PSR were flavol-3-ols, phenolic acids, flavonoid glycosides and hydrolysable tannin. Phenolic acid derivatives of PSR are caffeic acid glycoside dimmer and fabric acid derivate. Equally, procyanidin trimer type C, procyanidin dimer type B, procyanidin dimer and (E) catechin are the four flavan-3-ols identified from PSR. These four molecules exhibited a very strong antioxidant activity with the same λ_{max} of 280 nm. Procyanidin trimer type C showed [M-H]$^-$ ion of *m/z* 865 and fragments of *m/z* 849, 739, 713, 577, 407 and 425, and it was identified as proanthocyanidin trimer. Procyanidin dimer type B had a molecular ion of *m/z* 577 with fragments of *m/z* at 451, 425, 407, 289 and 245. The main fragments of procyanidin dimer were detected: *m/z* 289 and *m/z* 273 corresponding to catechin and afzelechin units, respectively. (+)-catechin presented at 23.9 min in HPLC profile with MS [M-H]$^-$ of *m/z* 289 and fragments of *m/z* 245, 205, 125 [37].

Quercetrin 3-O-rhamnoside, kaempferol 3-O-glucoside and kaempferol 3-O-rutinoside were characterized as flavonoid glycosides. These composites offered two maximum absorbances between 190 and 400 nm and could be related to flavonoid. Flavonoid has two maximum absorbances in ultraviolet region: 240–285 nm (band II) and 300–400 nm (band I) [51].

The MS/MS spectra of the reference substances quercetin, kaempferol 3---rutinoside-7-rhamnoside, and kaempferol 3-Oglucoside-7-rhamnoside showed a different ion [M - H - 146]$^-$ that visibly signify the removal of rhamnosyl moiety from the hydroxyl group of carbon 7 and confirmed approximately the same intensity as the aglycone [51].

Determination of the existence of biologically active O-prenylated umbelliferone derivatives, such as auraptene and umbelliprenin in ethanolic PS extracts, ultra high performance liquid chromatography (UHPLC) methodology with spectrophotometric (UV/Vis) detection, coupled with different extraction procedures, has been perfected by Fiorito *et al.* [10]. The highest concentration values recorded under short ultrasound-assisted conditions were 1.99 μg/g of dry extract and 6.53 μg/g for auraptene and umbelliprenin, respectively. The parent metabolite umbelliferone was also detected (0.67 μg/g).

In Table **2**, some of the most commonly used methods to identify and quantify phenolic compounds (chromatographic conditions; mobile phase and gradient, quantification and detection) from PS extracts are shown.

Table 2. Phenolic compounds characterization by analytical techniques reported in PS from different regions.

Source	Solvent Extraction	Analyte	Method	Quantification	Detection	Reference
Canada+ Brazil	Defatted with Hexane	Phenolic compounds (Free, esterified and insoluble-bound phenolic compounds)	HPLC-DAD-ESI-MS	47 phenolic compounds were found	Phenolic acids and Flavonoids (280nm) Anthocyanins (520nm)	[40]
Malaysia	Water	Phenolic acids	Liquid chromatography–mass spectrometry LC-MS	7 compounds were found	220 nm	[52]
China	Water	Phenolic acids	LC-MS/MS HPLC–DAD	Gallic acid 0.02-0.03 Ellagic acid 0.02-0.05 Punicalagin A et B (nd)	377 nm	[41]

(Table 2) cont.....

Source	Solvent Extraction	Analyte	Method	Quantification	Detection	Reference
Egypt	Water	Phenolic composition (mg/g)	HPLC	- Gallic acid 0.42-2.29 - Chlorogenic acid 0.04-0.73 -Caffeic acid 0.05-0.51 - Catechol 0.07-0.62 - Vanillic acid 0.03-0.21 - P-coumaric acid 0.01-0.05	220 nm	[42]
China	- Acetone (50%) (A) - Methanol (80%) (M) Acetone/ Methanol/Water	Phenolic acid composition Total soluble conjugated and insoluble bound phenolic acids (μmol/g)	HPLC	1.22-1.85	220 nm	[35]

Potential Antimicrobial Activity of PS

The study of Gaber *et al.* [53] was planned to evaluate the antibacterial activity of the DMSO, ethanol and methanol extracts of Egyptian and Taif (The Kingdom of Saudi Arabia) cultivars of PS extracts against five opportunistic pathogens namely *Staphylococcus aureus*, *Pseudomonas* sp. and *Bacillus* sp. (Gram positive bacteria), *Escherichia coli* and *Aeromonas hydrophila* (Gram negative bacteria). The inhibitory activity was found to be dose dependent.

In fact, the maximum antimicrobial activity for the PS extracts was evaluated at 60 mg/mL. *Escherichia coli* was reported to have considerable vulnerability against most of the extract, followed by *Staphylococcus aureus* and *Pseudomonas* sp.

The ethanolic extracts of Iranian PS were arranged and the antibacterial activity of extracts was determined by agar diffusion and micro-broth dilution methods against clinical isolates of *P. aeruginosa* and *S. aureus* strains [54]. PS showed inhibitory effects and the minimum inhibitory concentrations (MICs) of PS extracts were 25.0 mg/mL. In addition, the minimum bactericidal concentration ns (MBCs) of PS extracts were found to be 50 mg/mL.

According to Nozohour *et al.* [54], for all of the studied bacterial isolates, the

MICs and MBCs values for PS extract were higher than those of pomegranate peel extract ($P < 0.05$). From aqueous PS extracts, Tanveer *et al.* [55] evaluated the antibacterial activity in opposition to chosen pathogenic microorganisms: *Staphylococcus aureus*, *Pseudomonas aeruginosa*, *Escherichia coli*, and *Enterococcus feacalis* using disc inhibition zone technique and the results were evaluated with a commercial antibiotic (Amoxicillin). Lowest inhibition zone was distinguished for the seed aqueous extract of pomegranate. In fact, PS extract was three times inferior (21.12mm/7.13m) than amoxicillin [55].

PS Antioxidant Activity

The study of Ambigaipalan *et al.* [40] shows that phenolics from all PS fractions (free, ester, insoluble bound) are capable of scavenging 2,2-diphenyl-1-picrylhydrazyl (DPPH) and hydroxyl radicals. In this line, the total antioxidant aptitude of PS was resoluted by evaluating potential scavenging of the 2,2'-azinobis (3-ethylbenzothiazoline-6-sulphonate) radical cation ($ABTS^{\cdot+}$) and reported as micromoles of Trolox equivalents (TE). PS displayed ABTS radical cation scavenging activity with values of 11-18 µmol of Trolox equivalents/g sample. Under other conditions, the DPPH and hydroxyl radicals scavenging activities of PS were studied *via* an electron paramagnetic resonance (EPR) spectrometry.

DPPH radical scavenging activity of PS ranged around 0.09 and 0.43 mmol of Trolox/g defatted sample. Among all samples, PS ester had the maximum DPPH radical scavenging [40]. Singh *et al.* [34] described that at 100 ppm, ethyl acetate, methanol and water extracts of PS displayed 26.5, 23.2 and 39.6% DPPH radical scavenging activities, respectively.

In addition, He *et al.* [37] explained that subcritical water extraction of PS had elevated DPPH radical scavenging activity evaluated to ultrasound-assisted extraction.

Thitipramote *et al.* [36] investigated the antioxidant activities (DPPH and ABTS radical scavenging activities and ferric reducing antioxidant power (FRAP)) of Thailand PS that was extracted by three different solvents (water, 95% ethanol and 70% acetone) at ratio of 1:10 (w/v). Interestingly, the results showed that the furthermost ABTS activity was given from acetone seed extract (3.319 ± 0.016 mg TEAC/g) followed by PS with ethanol extract (1.420 ± 0.021 mg TEAC/g, $P < 0.05$).

In 2015, Basiri work was designed to estimate the consequence of solvents on extraction from PS and pomegranate defatted seed (PDS) and to measure the antioxidant properties [33]. Different solvents, including water, methanol, acetone, ethyl acetate and hexane were evaluated. The results illustrated that the peak of extraction efficiencies was for hexane and acetone solvents in extraction of seed and defatted seed, respectively. Radicals scavenging property and ferric reducing-antioxidant power of extracts were measured. Using the DPPH method, the antiradical activity (free radical-scavenging) of PS and PDS extracts was determined. The results recommend that the highest antiradical potential was demonstrated in methanol extracts of PS and PDS, on a µg analyte basis (EC_{50} (PS) = 0.14 and $EC_{50 (PDS)}$ = 0.19 µg/g). Reducing activity test confirmed that the methanol extracts of PS and PDS had the maximum reducing potency. The order of antioxidant capacity of PS and pomegranate defatted seed were established to be Methanol > Water > Acetone > Butanol > Ethyl acetate > Hexane.

Basiri [33] concluded that the methanol extract had elevated antioxidant effectiveness than seed and defatted seed extracts. The antioxidant activity of pomegranate was determined by FRAP assay. Post-hoc comparison using the Tukey HSD test showed the mean score of total antioxidant activity (TAA) for PS-juice (PSJ) (47 ± 5.5 mmol/L Fe^{+2}) was significantly ($P < 0.05$) different from PS and pomegranate juice (PJ) [33].

In 2012, Elfalleh *et al.* [28] demonstrated that Tunisian PS, antioxidants contents, expressed by DPPH (IC50), ABTS and FRAP, of both the aqueous and methanolic extracts were as follows: peel > flower > leaf > PS.

The bioactive compounds and antioxidant activities of juice, peel and seed of four Turkish genotypes pomegranate were evaluated by Orak *et al.* [45] in 2012. The DPPH scavenging activity values in peel extract (PE) were 23.4-fold superior than the juice extracts (JE), and the extracts of the seeds (SE) have 2.3-fold advanced scavenging activity than JE. The reducing power in PE was 4.7-fold elevated than SE and 10.5-fold higher than JE. The peak of metal-chelating capacity (37.22%) was shown in the peel, the lowest (7.151%) in SE.

Two *in vitro* assays established on DPPH and $ABTS^+$ radical scavenging capacities, respectively, were used to evaluate the antioxidant capacity of crude extract and various fractions of China PSR [37]. Radical scavenging competences of crude extract and fractions (ethyl ether phase fraction (F_a), ethyl acetate fraction (F_b), *n*-butanol phase fraction (F_c) and the residue after the fractionation with diverse solvents (F_d) against DPPH and $ABTS^+$ radicals were explored. Both of the techniques designates that the F_c possessed the strongest radical scavenging ability, with the inhibition of 92.75% against DPPH· (the concentration of F_c was 0.4 mg/mL) and 99.80% against $ABTS^+$ (the concentration of F_c was 0.2 mg/mL). The F_a exhibited inferior radical scavenging capacity than F_b with both of the ways. The F_d illustrated the weakest action, with the reticence of 30.48% against

DPPH (the concentration of F_d was 0.4 mg/mL) and 11.27% against ABTS$^+$ (the concentration of F_d was 0.2 mg/mL). The F_c verified the top radical scavenging aptitude since more quantities of phenolic compounds dissolved in *n*-butanol.

Using the bleaching of β-carotene, the antioxidant activity of Indian PS extracted by ethyl acetate, methanol and water, was measured [34]. At the concentration of 100 ppm, EtOAc, MeOH, and water extracts of PS demonstrate 39, 22, and 57% antioxidant activities, respectively. The presence of different extracts can hinder the extent of β-carotene bleaching by defusing the linoleate free radical and other free radicals produced in the system. Free radical scavenging potentials of PS extracts at different concentrations were experienced by the DPPH technique. At 100 ppm, EtOAc, MeOH, and water extracts of PS show 26.5, 23.2, and 39.6% free radical scavenging activities, respectively [34].

Two *in vitro* antioxidant assays, ferric reducing antioxidant power (FRAP) and Trolox equivalent antioxidant capacity (TEAC), were utilized to judge antioxidant capacity of six Georgian pomegranate cultivars [31]. Overall, the maximum antioxidant competence was found in leaves followed by peel, pulp and seed. TEAC values were lesser than the FRAP values, with the lowest being in PS (5.2 μM TE/g of FW) and the highest in leaf (13.7 μM TE/g of FW).

Antioxidant activity of the Indian ethanolic PS extract was measured by DPPH and hydrogen peroxide free radical scavenging method [56]. The extract demonstrated maximum scavenging activity *i.e.* 82.80±0.267 and 74.91±0.235 at 100 μg/mL by DPPH and hydrogen peroxide method, respectively as compared to the standard (ascorbic acid).

In the study of Sadeghi *et al.* [57] in 2012, the FRAP assay was employed and seed fraction FRAP values of six different cultivars of pomegranate in Iran were resoluted in an attempt to compare their differing antioxidant activities. Results showed that the extracts found from PS using a variety of solvents exhibited various levels of antioxidant activity.

In fact, PS antioxidant activity of six diverse cultivars of pomegranate in water extracts explained that the Sour white peel cultivar has the maximum of FRAP value (3.45±0.85 μM) and the Agha Mohamad Ali cultivar has the lowest value (2.76±0.76 μM); ethanolic extract of the seeds illustrated that Sour white peel and Black peel cultivars have the highest (3.88±1.31 μM) and lowest (1.62±0.47 μM) antioxidant activity, respectively [58]. Surveswaran *et al.* [58] in 2007 have reported a systematic evaluation of natural phenolic antioxidants from 133 medicinal plants in India. In their description, the PS and the pericarp demonstrate FRAP values around 0.94 and 19.22 μmol Trolox per gram dry weight. The antioxidant activities of PP reported in the literature were illustrated in Table **3**.

Table 3. Antioxidant capacity of some PS extracts.

Extraction	Method	Antioxidant activity	Reference
Water	DPPH (mg TEAC /g extract)	0.034	[36]
	ABTS (mg TEAC /g extract)	0.366	
	FRAP (mg TEAC /g extract)	0.560	
	DPPH (%) at 200µg/mL	53.98	[52]
	DPPH (EC_{50} (mg/mL))	159.9	
	DPPH (%)	35-80	[42]
	β-Carotene (%)	32-70	
	DPPH (EC_{50} (mg/mL))	0.3-0.32	[33]
	Ferric reduction power (μMol Fe^{2+}/L)	207.6-217.3	
	Vitamin C	(-)	[43]
	DPPH (EC_{50} (mg/mL))	45.05	[28]
	$ABTS^+$ (TEAC mmol/100 g DW)	0.76	
	Reducing power (EC_{50} (mg/mL))	321.15	
	DPPH (TEAC mmol/100 g DW)	483.6-4187	[37]
	ABTS (TEAC mmol/100 g DW)	494-2692	
	ABTS (only water extract)	9 compounds were found	
	DPPH (EC_{50} (µg/mL))	2.577	[44]
	DPPH (EC_{50} (mg/mL))	1.136-2.518	[45]
	β-Carotene bleaching assay (%)	Between 70 and 80%	
	Reducing power (700nm)	Less than 0.5	
	Metal chelating capacity (%)	Between 5 and 12%	
	DPPH (g/g)	Dry seeds 6.5 Wet seeds 6	[46]

(Table 3) cont.....

Extraction	Method	Antioxidant activity	Reference
Acetone	DPPH (mg TEAC /g extract)	0.512	[36]
	ABTS (mg TEAC /g extract)	3.320	
	FRAP (mg TEAC /g extract)	1.903	
	DPPH (EC_{50} (mg/mL))	0.24-0.34	[33]
	FRAP (μMol Fe^{2+}/L)	40.7-155.1	
	DPPH (TEAC mmol/100 g DW)	1412.2-3695	[37]
	ABTS (TEAC mmol/100 g DW)	475.8-1138	
	DPPH (EC_{50} (mg/mL))	6.4-13.1	[35]
	ABTS (μmol TE/g)	8.5-17.8	
	FRAP assay (μmol TE/g)	6.4-8.6	
Methanol	DPPH (EC_{50} (mg/mL))	0.15-0.19	[33]
	Ferric reduction power (μMol Fe^{2+}/L)	557-721.8	
	DPPH (EC_{50} (μg/mL))	21	[28]
	$ABTS^+$ (TEAC mmol/100 g DW)	1.10	
	Reducing power (EC_{50} (μg/mL))	337.84	
	DPPH (TEAC mmol/100 g DW)	521-4265.7	[37]
	ABTS (TEAC mmol/100 g DW)	474.7-1690	
	DPPH (EC_{50} (μg/mL))	11.2-19.8	[35]
	ABTS (μmol TE/g)	7.4-12.9	
	FRAP assay (μmol TE/g)	6-7.8	

(Table 3) cont.....

Extraction	Method	Antioxidant activity	Reference
	DPPH (EC_{50} (mg/mL))	0.461-0.502	[45]
	β-Carotene bleaching assay (%)	Between 60 and 70%	
	Reducing power (700 nm)	Near 0.5	
	Metal chelating capacity (%)	Between 2 and 5%	
Ethyl acetate	DPPH (EC_{50} (mg/mL))	1.83-2.01	[33]
	Ferric reduction power (μMol Fe^{2+}/L)	22.2-35.1	
Ethanol 95%	DPPH (mg TEAC /g extract)	0.066	[36]
	ABTS (mg TEAC /g extract)	1.420	
	FRAP (mg TEAC /g extract)	0.935	
Ethanol 80%	Antioxidant activity (%) β-carotene bleaching test	26- 54	[36]
Ethanol 70%	FRAP (mmol Fe+2/L)	20	[43]
	Vitamin C	(+)	[37]
	DPPH (TEAC mmol/100 g DW)	1695-3391.3	
	ABTS (TEAC mmol/100 g DW)	493.8-1084.7	
	DPPH (EC_{50} (μg/mL))	1.324	[44]
	DPPH (EC_{50} (mg/mL))	0.544-0.64	[45]
	β-Carotene bleaching assay (%)	Between 60 and 75%	
	Reducing power (700 nm)	Between 0.15 and 0.4	
	Metal chelating capacity (%)	Between 1 and 2%	

(Table 3) cont.....

Extraction	Method	Antioxidant activity	Reference
Hexane	ABTS (µmol Trolox/g)	11-18.3	[36]
	DPPH (mmol Trolox/g)	0.2-0.43	
	Hydroxyl scavenging activity (µmol GAE / g)	17.5-109	
	Metal chelating ability (µmol EDTA/g)	0.11-3.03	
	DPPH (EC_{50} (mg/mL))	3.88-4.23	[33]
	Ferric reduction power (µMol Fe^{2+}/L)	6.4-9.9	
	Tocochromanols (mg/100 g Dry)	98.35	[59]
	FRAP Assay (µM TE/ g FW)	7.8-9 Lipophilic 12.5-19 Hydrophilic	[31]
	ABTS (µM TE/ g FW)	5.2-6 Lipophilic 7.5-9 Hydrophilic	
	Vitamin E (Tocopherols α β γ δ) (mg/100g)	263-290.3	
Butanol	DPPH (EC50 (mg/mL))	1.69-1.77	[33]
	Ferric reduction power (µMol Fe^{2+}/L)	59.3-60.5	
Chloroform	Vitamin C	(+)	[43]

Biopreservation of Food and Agri-food Products Formulated with PS Extracts

Meat Products

Goat Meat

Effects of salt, kinnow and pomegranate fruit by-product powders on color and oxidative stability of raw ground goat meat stored at 4 ± 1 °C was evaluated by Devatkal and Naveena [60] in 2010. In this study, five treatments: control (only meat), MS (meat + 2% salt), KRP (meat + 2% salt + 2% kinnow rind powder), PRP (meat + 2% salt + 2% pomegranate rind powder) and PSP (meat + 2% salt + 2% PS powder) were evaluated. During 6 days of refrigerated storage, the Hunter Lab L value (L^*) increased in control and remained unchanged in others during 12 days of storage, while, redness scores declined and yellowness showed inconsistent changes. For lipid oxidation, throughout storage, TBARS (thiobarbituric acid reactive substances) values were higher ($P < 0.05$) in MS followed by control and KRP samples compared to PRP and PSP samples. The PSP treated samples showed lowest TBARS values than others. Percent reduction of TBARS values was highest in PSP (443%) followed by PRP (227%) and KRP (123%). Generally, the overall antioxidant effect was in the order of PSP > PRP > KRP > control > MS [60].

In 2010, Devatkal *et al.* [61] used kinnow rind powder (KRP), pomegranate rind powder (PRP) and PS powder (PSP) extracts at 1% in goat meat patties. This study showed that $L*$ value significantly ($P < 0.05$) lowered in PRP followed by PSP and KRP patties during 12 days in refrigerated storage. Sensory evaluation indicated no significant differences ($P > 0.05$) among patties. Further, a significant ($P < 0.05$) reduction in TBARS values (lipid oxidation) during storage of goat meat patties was observed in PRP, PSP and KRP as compared to control patties [61].

According to Narsaiah *et al.* [62], tenderization of goat meat with PS powder improved the textural properties marginally with slight adverse color change and taste. Treated samples got lower score for color in sensory evaluation and there was adverse effect on taste of treated meat. Blade tenderization and 4% PSP proved better for tenderization and were compared with control and 0.2% papain in goat meat chunks. The cooked samples treated with papain and blade incisions got better sensory scores and required lesser shear force compared to 4% PSP and control. Overall, meat treatment with papain was better in terms of sensory attributes improvement followed by that with blade incision, and PS powder might be considered for mixture with other spices to marinate goat meat mainly for its beneficial effects [62].

Pork Meat

In 2013, Qin *et al.* [63] evaluated the antioxidant potential of pomegranate rind powder extract (PRP), pomegranate juice (PJ) and PS powder extract (PSP), at 20 mg equivalent PP extract phenolics/100 g meat in raw ground pork meat stored at $4 \pm 1°C$ for 12 days. The standard plate count in the PRP group was significantly ($P < 0.05$) lower than that in all other groups. PRP significantly ($P < 0.05$) reduced lipid oxidation compared to PJ and PSP. Samples with antioxidants had significantly ($P < 0.05$) decreased peroxide formation than control groups. Antioxidant effectiveness was in the order: BHT (Butylated hydroxytoluene) > PRP > PJ > PSP > control. $L*$ value was lowered by the addition of PRP. The overall acceptability scores of PRP, PJ, PSP and BHT treated samples were higher than that of control samples. The results indicated the potential of natural functional ingredients to enhance the quality of raw ground pork meat [63].

Beef and Poultry Meat

In 2014, Keşkekoğlu and Üren [64] studied the cooked beef and chicken meatballs with a 0.5% (w/w) PS extract using four different cooking methods (oven roasting, pan cooking, charcoal-barbecue and deep-fat frying) and observed six heterocyclic aromatic amines; 2-amino-3-methylimidazo[4,5-f]quinoline (IQ); 2-amino-3,8-dimethylimidazo [4,5-f]quinoxaline (MeIQx); 2-amino-1-meth-l-6-phenylimidazo [4,5-b] pyridine (PhIP); 9H-pyrido [3,4-b] indole (norharman) and 1-methyl-9H-pyrido [3,4-b] indole (harman). In the beef meatballs, the highest inhibitory effects of PS extract on heterocyclic aromatic amines formation were 68% for PhIP, 24% for norharman, 18% for harman, 45% for IQ and 57% for MeIQx. Total heterocyclic aromatic amine formation was reduced by 39% and 46% in beef meatballs cooked by charcoal-barbecue and deep-fat frying, respectively. In the chicken meatballs, the highest inhibitory effects were 75% for PhIP, 57% for norharman, 28% for harman, 46% for IQ and 49% for MeIQx. When the PS extract was added to the chicken meatballs cooked by deep-fat frying, the total heterocyclic aromatic amine formation was inhibited by 49%, in contrast to the total heterocyclic aromatic amine contents after oven roasting increased by 70%.

Kaur *et al.* [65] in 2015 evaluated the effect of PS powder on the quality characteristics of chicken nuggets during refrigerated storage. The products were developed by incorporating optimum level of PS powder (3%) and was aerobically packaged in low-density polyethylene pouches and assessed for various storage quality parameters under refrigerated (4 ± 1°C) conditions for 21 days of storage. The products were evaluated for various physico-chemical, microbiological and sensory parameters at regular intervals of 0, 7, 14 and 21 days. In this study, a significant ($P < 0.05$) effect of PS powder was observed on the pH and TBARS (mg malondialdehyde/kg) values of the chicken nuggets. A significant ($P < 0.05$) effect was also observed on the microbiological characteristics: for total plate count, psychrophilic count as well as yeast and mould count during the period of storage. Coliforms were not detected throughout the period of storage.

Significantly ($P < 0.05$) higher scores were observed for various sensory parameters of the products incorporated with PS at 3%. This concentration could be successfully improved the oxidative stability and storage quality of the products during chilled (4 ± 1°C) storage and may be commercially exploited to improve the storage quality of muscle foods without adversely affecting the sensory quality of the products [65].

Aquatic Meat Product

In 2011, Özalp Özen *et al.* [66] investigated the effect of PS extract (PSE) and grape seed extract (GSE) addition to chub mackerel minced muscle on lipid oxidation during frozen storage.

Each extract was added to minced fish muscle at 2% concentration and then stored at −18°C for 3 months. The effect of plant dietary fibers to control lipid oxidation was compared with untreated samples (control). Formation of lipid hydroperoxides and TBARS was significantly inhibited by PSE and GSE addition when compared with control. Both extracts considerably retarded lipid oxidation according to the results of TBARS. A significant reduction of $L*$ (lightness), $a*$ (redness) and $b*$ (yellowness) values were detected during frozen storage. PSE added samples had the highest redness and the lowest lightness and yellowness. However, samples with GSE showed the lowest redness and highest yellowness and h° (hue angle) values. The results from this study suggest that PSE is a very effective inhibitor of primary and secondary oxidation products in minced fish muscle and have a potential as a natural antioxidant to control lipid oxidation during frozen storage of fatty fish [66].

Pasta

According to Dib in 2018, incorporating PS powder (PSP) at low levels enhanced the nutritional quality of pasta without a significant adverse effect on its cooking, textural and sensory properties [67]. In this line, PSP supplementation to gluten-free (GF) sheeted pasta on cooking properties, sensory characteristics and antioxidants properties using TLC-DPPH test were studied. Five levels of PSP were used (2.5, 5.0, 7.5, 10.0 and 12.5%) on formula replacement basis. Antioxidant potential of GF pasta increased with the addition of PSP.

The GF pasta without additives and with 2.5% concentration of PSP did not reveal ability to scavenge free radicals. The highest aforementioned activity was observed for crude PS extract followed by GF pasta with 12.5, 10.0, 7.5 and 5.0% addition of PSP. The total dietary fibers content of pasta increased from 5.68 to 14.80 g/100 g with the increase in the incorporation of PSP from 0 to 12.5%. The results revealed that cooking loss of gluten-free pasta increased from 9.09 to 10.18%, whereas pasta firmness decreased from 381.43 to 366.30 N, upon incorporation of PSP. On the other hand, PSP decreased the lightness of the pasta from 82.26 to 57.27. Sensory analysis suggested that control pasta (without PSP) and pasta supplemented with low levels of PSP have the most acceptable quality [67].

Yoghurts

The effect of phenolic compounds extracted PS on characteristics of strained yoghurts produced from sheep milk was investigated by Ersöz *et al.* [68]. Firstly, phenolic compounds were extracted PS and its amounts for strained yoghurts were determined by estimating their phenolic activities. Also antimicrobial activity of PS extracts was determined. On the 1st, 7th, and 14th day of storage, chemical analyses as dry matter, protein, acidity, pH, proteolysis and peroxide value, microbiological analyses such as enumeration of *Lactobacillus delbrueckii* ssp. *bulgaricus* and *Streptococcus thermophilus* and sensory analysis were conducted on strained yoghurt samples. According to analysis results, while addition of phenolic compounds affects chemical and microbiological properties of strained yoghurt positively, sensory quality was affected negatively [68].

Van Nieuwenhove *et al.* [16] characterized and employed PS powder for the bioactive lipid enrichment of yoghurt. Higher scavenging activity and phenolic content were determined for PS. Compared to the control, yoghurt enriched with 0.5% (w/v) PS flour showed similar nutritional and pH values, but higher antioxidant activities, desirable seed oil fatty acid and conjugated linolenic acids (CLnA) contents, and lower atherogenicity indexes. Bio-fortified yoghurt showed high overall acceptability. This study focused for the first time on the feasibility of producing CLnA-enriched dairy foods using non-conventional plant seeds or waste seeds [16].

Animal Feeding

Saki *et al.* [69] investigated the effect of supplementation on different levels of PS pulp (PSP) on performance and blood parameters in laying hens. A total of 96 layers hens (Hy-line W-36) at 24 weeks of age were randomly assigned into 4 treatments including 0 (control), 5, 10 and 15% of PSP, with 4 replicates containing 6 layers in each. Results showed no significant effect of PSP levels on feed intake, egg mass, egg weight, feed conversion ratio and body weight gain. Supplementation of PSP at 5% increased egg production and is significantly different ($P < 0.05$) to 15% PSP but not in control group, suggesting negative effect of high level PSP in layers diet. Haugh unit, yolk and albumen indexes as well as eggshell weight, eggshell ratio and breaking strength were not significantly ($P > 0.05$) affected by dietary PSP. There was no significant effect of PSP ($P > 0.05$) on serum triglycerides, high-density lipoprotein and total antioxidant. In contrast, the serum malondialdehyde (MDA) was significantly ($P < 0.05$) increased using 5% PSP in diet with a notable ($P < 0.05$) increase of cholesterol in all inclusion levels of PSP compared to the control. The result showed that supplementation of PSP up to 15% improves egg production but higher concentration has detrimental effects on laying performance. Equally, the addition of PSP increased ($P < 0.05$) cholesterol in the layers of blood [69].

The effect of level of dried PS pulp (PSP) in the diet of goat kids on meat quality and fatty acid profiles of intramuscular and subcutaneous fat was studied by Emami *et al.* [18] in 2015a.

Thirty two *Mahabadi* goat kids were randomly allocated to four dietary treatments: without PSP (control), containing 50 g PSP/kg DM (PSP5), containing 100 g PSP/Kg DM (PSP10) and containing 150 g PSP/Kg DM (PSP15). At the end of the 84-day feeding trial, the goat kids were slaughtered and *M. longissimus lumborum* (LL) and subcutaneous adipose tissues were sampled. Addition of PSP linearly increased ($P = 0.01$) fat content and decreased ($P < 0.01$) shear force, drip loss, total aerobic bacterial count and lipid oxidation of LL muscle. Feeding PSP diets linearly increased the concentrations of C18:2 n-6 ($P < 0.01$), C18:3 n-3 ($P < 0.001$), n-6 polyunsaturated fatty acids (PUFA; $P < 0.01$) and n-3 PUFA ($P < 0.001$), and decreased ($P < 0.05$) the ratio of n-6/n-3 in both muscle and adipose tissues. A linear increase was observed in vaccenic acid (VA, $P < 0.01$), conjugated linoleic acid (CLA, $P < 0.001$) and punicic acid (PUA, $P < 0.001$) concentration in subcutaneous and intramuscular fat, with increasing PSP level in diet. Emami *et al*. [18] concluded that PSP supplementation of goat kids' diet up to 150 g/Kg DM can improve the nutritional and functional properties of meat.

Later, these authors investigated the effect of partial replacing of cereal grains of diet with PSP on performance, nutrient digestibility and antioxidant capacity of fattening *Mahabadi* goat kids [19]. Thirty two *Mahabadi* male goat kids, 4–5 months of age and 16.5 ± 2.8 Kg body weight (BW) were assigned to four dietary treatments: (1) diet without PSP (control), (2) diet containing 5% of PSP (PSP5), (3) diet containing 10% of PSP (PSP10) and (4) diet containing 15% of PSP (PSP15) (DM basis). The goat kids were slaughtered after 84 days of feeding trial and antioxidant capacity was measured in the liver and LL muscle samples. Dry matter intake (DMI), average daily gain (ADG) and feed conversion ratio (FCR) were not affected by diets ($P > 0.05$). Feed cost per Kg of hot and cold carcass weight decreased with increasing levels of PSP in diet ($P < 0.05$). Addition of PSP to diet decreased kidney fat ($P < 0.05$) and tended to increase ether extract (EE) apparent digestibility ($P = 0.07$). The LL muscle ($P < 0.05$), liver ($P = 0.08$) and plasma ($P < 0.05$) samples from goat kids fed PSP15 displayed a greater antioxidant capacity than goat kids fed control diet. No significant difference was found in glutathione peroxidase (GSH-Px) activity among the groups ($P > 0.05$), but the MDA content of the LL muscle, liver and plasma decreased ($P < 0.05$) in PSP15 group when compared with control group. These authors indicated that partial replacing of dietary cereal grains with PSP did not affect growth performance, carcass traits and nutrient digestibility, while decreased the cost of meat production and improved the antioxidant capacity of goat kids [19].

In other research work, these authors studied the effect of PSP on thirty-two *Mahabadi* goat male kids [20]. Different levels of PSP: (1) diet without PSP (control), (2) diet containing 5% PSP (PSP5), (3) diet containing 10% PSP (PSP10) and (4) diet containing 15% PSP (PSP15) were evaluated. The goat kids were slaughtered at the end of the study and the LL was sampled. The TBARS values of both raw and cooked meat decreased ($P < 0.0001$) by increasing of PSP levels in the diet. The meat of goat kids fed PSP15 showed higher $a*$ and $C*$ values ($P < 0.01$) and lower $H*$ and $b*$ values ($P < 0.001$), than goat kids fed with control diet. The results of Emami *et al.* study [20] indicated that replacing barley and corn grains with PSP in the diet may improve the color and lipid stability of goat kid meat.

The effects of feeding PSP on milk yield, milk composition, fatty acid profiles of milk fat and blood metabolites were examined by Modaresi [70] in 2011. During a pretrial period, 27 multiparous southern Khorasan (Iran) cross-bred goats were fed with a similar diet and dry matter intake, milk yield and milk composition were recorded. After adaptation and based on pretrial records, the goats were randomly assigned to 1 of 3 experimental diets and were housed in individual stalls. Experimental diets included 0, 6 or 12% of PSP (dry matter basis) and were fed as total mixed rations *ad libitum* for a 45-day period. Diets were formulated to be isonitrogenous and isocaloric. Supplementation of PSP did not affect dry matter intake or average daily gain of goats. Milk yield was also not affected by inclusion of PSP in the diet. Milk fat concentration of goats fed diets with 6 and 12% PSP increased, but milk fat yield, milk protein concentration and milk solids-not-fat concentration of goats were not affected by diets. Furthermore, it was observed that feeding PSP did not affect blood glucose, cholesterol, urea N, triglyceride or lipoproteins. Feeding goats with a diet containing 12% PSP modified the milk fatty acid profile, including conjugated linoleic, punicic and vaccenic acids [70].

In 2015, Razzaghi *et al.* [71] compared three underused agro-industrial by-products which were PS pulp (PSP), pistachio hulls (PH) and tomato pomace (TP) available in dry areas with respect to their potentiality influence on ruminal fermentation, performance and milk fatty acid (FA) profile of dairy goats. Eight multiparous lactating Saanen goats were randomly assigned to a 4 × 4 Latin square design with 4 dietary treatments over 21-day periods. The dietary treatments were: control diet (CON) and diets containing PSP (120 g/Kg DM), PH (240 g/Kg DM) or TP (240 g/Kg DM), substituting for wheat bran of the control diet. All diets were kept isoenergetic and isonitrogenous with forage to concentrate ratio of 45:55 (DM basis). The most abundant FA in the lipids of PSP and TP were c9,t11,c13-18:3 and c9,c12-18:2, respectively, while PH was rich in c9-C18:1 and phenolic compounds. No consistent treatment effects were observed on DM intake, milk yield, milk fat, protein and lactose yields. PS pulp increased ($P < 0.01$) milk fat concentration compared with CON and PH diets and protein concentration of the milk samples obtained from animals fed by PH diet was highest among the treatments. There was no diet effect on ruminal pH, while rumen ammonia-N concentrations, volatile fatty acid concentrations and, to a lesser extent, molar proportion of acetate were decreased following PH diet feeding. Blood cholesterol concentration increased ($P < 0.01$) with PSP and TP diets. The blood urea N concentration decreased ($P < 0.05$) when PH diet was fed. Feeding TP diet decreased ($P < 0.01$) 16:0 and tended ($P < 0.10$) to increase c9-18:1 proportion in milk fat in comparison to the other diets. Inclusion of all by-products increased ($P < 0.01$) t11-18:1 (2-fold) and total conjugated linoleic acids (CLA, 5 to 6-fold) in milk fat compared to CON diet. In addition, concentrations of c9, c12,c 15-18:3 and total polyunsaturated FA (PUFA) in milk fat component were highest ($P < 0.01$) in the milk samples of animals fed by PSP diet [71].

CONCLUSION

Pomegranate consumption has grown tremendously due to its reported health benefits. Rich sources of several high value compounds with potential beneficial, PS were successfully used in food and agri-food products biopreservation. Although, PS extracts are generally regarded as safe, further research is needed to determine their safe limits. Thus, nutritional and toxicological studies (*in vitro/in vivo*) are strongly recommended to be carried out in the near future to ascertain the safe edible use of these natural sources. This is the most important point because the food industry is rejecting synthetic antioxidants on the basis of negative health-related issues; thus, while accepting new natural antioxidants, these must be analyzed for the same health-related consequences.

FUTURE PERSPECTIVES

There has been a renaissance in the study of bioactive compounds and the interest in PS phenolics is intense. The usual procedure now encompasses a high-performance separation technique in combination with diode array detection or mass spectrometry. Further instrument sophistication in coupling several systems such as multidimensional chromatography with NMR and MS in series is already occurring. The prediction of the future for promising approaches involves the application of HPLC with ESI time of flight mass spectrometers and ESI FT ion cyclotron resonance mass spectrometers. An increased emphasis on microcapillary columns with nanotechnology ESI systems driven partly by environmental issues seems inevitable.

On the other hand, although, PS extracts are generally regarded as safe, further research is needed to determine their safe limits. Thus, nutritional and toxicological studies (*in vitro/in vivo*) must be done to ascertain the safe edible use of these natural sources. This is the most important point because the food industry is rejecting synthetic antioxidants on the basis of negative health-related issues; thus, while accepting new natural antioxidants, these must be analyzed for the same health-related consequences.

CONSENT FOR PUBLICATION

Not applicable.

CONFLICT OF INTEREST

The author(s) confirms that there is no conflict of interest.

ACKNOWLEDGEMENTS

Declared none.

REFERENCES

[1]   Chai TT, Tan YN, Ee KY, Xiao J, Wong FC. Seeds, fermented foods, and agricultural by-products as sources of plant-derived antibacterial peptides. Crit Rev Food Sci Nutr 2019; 59(sup1): S162-77.
         [http://dx.doi.org/10.1080/10408398.2018.1561418] [PMID: 30663883]

[2]   Sharma L, Saxena A, Maity T. Trends in the manufacture of coatings in the postharvest conservation of fruits and vegetables. Polymers for Agri-Food Applications 2019; pp. 355-75.
         [http://dx.doi.org/10.1007/978-3-030-19416-1_18]

[3]   Chhikara N, Kushwaha K, Sharma P, Gat Y, Panghal A. Bioactive compounds of beetroot and utilization in food processing industry: A critical review. Food Chem 2019; 272: 192-200.
         [http://dx.doi.org/10.1016/j.foodchem.2018.08.022] [PMID: 30309532]

[4]   Ramos M, Burgos N, Barnard A, *et al. Agaricus bisporus* and its by-products as a source of valuable extracts and bioactive compounds. Food Chem 2019; 292: 176-87.
         [http://dx.doi.org/10.1016/j.foodchem.2019.04.035] [PMID: 31054663]

[5]   Ben Salah H, Smaoui S, Abdennabi R, Allouche N. LC-ESI-MS/MS phenolic profile of *Volutaria lippii* (L.) *Cass*. Extracts and Evaluation of their *In Vitro* Antioxidant, Antiacetylcholinesterase, Antidiabetic, and Antibacterial Activities. Evid Based Complementary Altern Med 2019; 2019: p. 13.
         [http://dx.doi.org/10.1088/0031-9155/43/10/028] [PMID: 9814537]

[6]   Fourati M, Smaoui S, Ennouri K, *et al.* Multiresponse optimization of pomegranate peel extraction by statistical *versus* Artificial Intelligence: Predictive Approach for Foodborne Bacterial Pathogen Inactivation. Evid Based Complementary Altern Med. 2019; p. 18.
         [PMID: 1542615]

[7]   Fourati M, Smaoui S, Ben Hlima H, *et al.* Synchronised interrelationship between lipid/protein oxidation analysis and sensory attributes in refrigerated minced beef meat formulated with *Punica granatum* peel extract.. IJFST 2019.
         [http://dx.doi.org/https://doi.org/10.1111/ijfs.14398]

[8]   Ben Hlima H, Bohli T, Kraiem M, *et al.* Combined effect of *Spirulina platensis* and *Punica granatum* peel Extracts: Phytochemical content and antiphytophatogenic Activity. Appl Sci (Basel) 2019; 9: 5475-85.
[http://dx.doi.org/10.3390/app9245475]

[9]   Smaoui S, Hlima HB, Mtibaa AC, *et al.* Pomegranate peel as phenolic compounds source: Advanced analytical strategies and practical use in meat products. Meat Sci 2019; 158:107914.
[http://dx.doi.org/10.1016/j.meatsci.2019.107914] [PMID: 31437671]

[10]   Fiorito S, Ianni F, Preziuso F, *et al.* UHPLC-UV/Vis quantitative analysis of hydroxylated and *O*-prenylated coumarins in pomegranate seed extracts. Molecules 2019; 24(10): 1963.
[http://dx.doi.org/10.3390/molecules24101963] [PMID: 31121819]

[11]   Peng Y. Comparative analysis of the biological components of pomegranate seed from different cultivars. Int J Food Prop 2019; 22: 784-94.
[http://dx.doi.org/10.1080/10942912.2019.1609028]

[12]   Mathon C, Chater JM, Green A, *et al.* Quantification of punicalagins in commercial preparations and pomegranate cultivars, by liquid chromatography-mass spectrometry. J Sci Food Agric 2019; 99(8): 4036-42.
[http://dx.doi.org/10.1002/jsfa.9631] [PMID: 30729530]

[13]   Russo M, Cacciola F, Arena K, *et al.* Characterization of the polyphenolic fraction of pomegranate samples by comprehensive two-dimensional liquid chromatography coupled to mass spectrometry detection. Nat Prod Res 2020; 34(1): 39-45.
[http://dx.doi.org/10.1080/14786419.2018.1561690] [PMID: 30691301]

[14]   Amiryousefi MR, Mohebbi M, Tehranifar A. Pomegranate seed clustering by machine vision. Food Sci Nutr 2017; 6(1): 18-26.
[http://dx.doi.org/10.1002/fsn3.475] [PMID: 29387357]

[15]   Koca I, Tekguler B, Yilmaz VA, Hasbay I, Koca AF. The use of grape, pomegranate and rosehip seed flours in Turkish noodle (erişte) production. J Food Process Preserv 2018; 42: e13343.
[http://dx.doi.org/10.1111/jfpp.13343]

[16]   Van Nieuwenhove CP, Moyano A, Castro-Gómez P, *et al.* Comparative study of pomegranate and jacaranda seeds as functional components for the conjugated linolenic acid enrichment of yogurt. Lebensm Wiss Technol 2019; 111: 401-7.
[http://dx.doi.org/10.1016/j.lwt.2019.05.045]

[17]   Martínez L, Castillo J, Ros G, Nieto G. Antioxidant and antimicrobial activity of rosemary, pomegranate and olive extracts in fish patties. Antioxidants 2019; 8(4): E86.
[http://dx.doi.org/10.3390/antiox8040086] [PMID: 30987153]

[18]   Emami A, Nasri MF, Ganjkhanlou M, Rashidi L, Zali A. Dietary pomegranate seed pulp increases conjugated-linoleic and-linolenic acids in muscle and adipose tissues of kid. Anim Feed Sci Technol 2015; 209: 79-89.
[http://dx.doi.org/10.1016/j.anifeedsci.2015.08.009]

[19]   Emami A, Ganjkhanlou M, Nasri MF, Zali A, Rashidi L. Pomegranate seed pulp as a novel replacement of dietary cereal grains for kids. Small Rumin Res 2015; 123: 238-45.
[http://dx.doi.org/10.1016/j.smallrumres.2014.12.001]

[20]   Emami A, Nasri MH, Ganjkhanlou M, Zali A, Rashidi L. Effects of dietary pomegranate seed pulp on oxidative stability of kid meat. Meat Sci 2015; 104: 14-9.
[http://dx.doi.org/10.1016/j.meatsci.2015.01.016] [PMID: 25681560]

[21]   Watson L, Dallwitz MJ. The families of flowering plants: Descriptions, illustrations, identification, and information retrieval 1992. http:// delta-intkey.com/angio/www/punicace.htm

[22]   Linnaeus C. Species Plantarum. Stockholm, Sweden 1753; Vol. I: p. 472.

[23]   Moreno PM, Valero RM. El granado. Mundi-Prensa 1992.

[24]   Levin GM. Pomegranate (*Punica granatum*) plant genetic resources in Turkmenistan. Bulletin des Ressources Phytogenetiques (IPGRI/FAO); Noticiario de Recursos Fitogeneticos (IPGRI/FAO) 1994.

[25]   da Silva JAT, Rana TS, Narzary D, Verma N, Meshram DT, Ranade SA. Pomegranate biology and biotechnology: A review. Sci Hortic (Amsterdam) 2003; 160: 85-107.
[http://dx.doi.org/10.1016/j.scienta.2013.05.017]

[26]   Bchir B, Besbes S, Karoui R, Attia H, Paquot M, Blecker C. Effect of air-drying conditions on physico-chemical properties of osmotically pre-treated pomegranate seeds. Food Bioprocess Technol 2012; 5: 1840-52.
[http://dx.doi.org/10.1007/s11947-010-0469-3]

[27]   Kulkarni AP, Aradhya SM. Chemical changes and antioxidant activity in pomegranate arils during fruit development. Food Chem 2005; 93: 319-24.
[http://dx.doi.org/10.1016/j.foodchem.2004.09.029]

[28]   Elfalleh W, Hannachi H, Tlili N, Yahia Y, Nasri N, Ferchichi A. Total phenolic contents and antioxidant activities of pomegranate peel, seed, leaf and flower. J Med Plants Res 2012; 6: 4724-30.
[http://dx.doi.org/10.5897/JMPR11.995]

[29]   Gözlekçi S, Saraçoğlu O, Onursal E, Özgen M. Total phenolic distribution of juice, peel, and seed extracts of four pomegranate cultivars. Pharmacogn Mag 2011; 7(26): 161-4.
[http://dx.doi.org/10.4103/0973-1296.80681] [PMID: 21716925]

[30]   Kalaycıoğlu Z, Erim FB. Total phenolic contents, antioxidant activities, and bioactive ingredients of juices from pomegranate cultivars worldwide. Food Chem 2017; 221: 496-507.
[http://dx.doi.org/10.1016/j.foodchem.2016.10.084] [PMID: 27979233]

[31]   Pande G, Akoh CC. Antioxidant capacity and lipid characterization of six Georgia-grown pomegranate cultivars. J Agric Food Chem 2009; 57(20): 9427-36.
[http://dx.doi.org/10.1021/jf901880p] [PMID: 19743855]

[32]   Derakhshan Z, Ferrante M, Tadi M, *et al.* Antioxidant activity and total phenolic content of ethanolic extract of pomegranate peels, juice and seeds. Food Chem Toxicol 2018; 114: 108-11.
[http://dx.doi.org/10.1016/j.fct.2018.02.023] [PMID: 29448088]

[33]   Basiri S. Evaluation of antioxidant and antiradical properties of Pomegranate (*Punica granatum* L.) seed and defatted seed extracts. J Food Sci Technol 2015; 52(2): 1117-23.
[http://dx.doi.org/10.1007/s13197-013-1102-z] [PMID: 25694727]

[34]   Singh RP, Chidambara Murthy KN, Jayaprakasha GK. Studies on the antioxidant activity of pomegranate (*Punica granatum*) peel and seed extracts using *in vitro* models. J Agric Food Chem 2002; 50(1): 81-6.
[http://dx.doi.org/10.1021/jf010865b] [PMID: 11754547]

[35]   Jing P, Ye T, Shi H, *et al.* Antioxidant properties and phytochemical composition of China-grown pomegranate seeds. Food Chem 2012; 132(3): 1457-64.
[http://dx.doi.org/10.1016/j.foodchem.2011.12.002] [PMID: 29243636]

[36]   Thitipramote N, Maisakun T, Chomchuen C, *et al.* Bioactive compounds and antioxidant activities from pomegranate peel and seed extracts. FABJ 2019; 7: 152-61.

[37]   He L, Zhang X, Xu H, *et al.* Subcritical water extraction of phenolic compounds from pomegranate (*Punica granatum* L.) seed residues and investigation into their antioxidant activities with HPLC–ABTS+ assay. Food Bioprod Process 2012; 90: 215-23.
[http://dx.doi.org/10.1016/j.fbp.2011.03.003]

[38]   Anahita A, Asmah R, Fauziah O. Evaluation of total phenolic content, total antioxidant activity, and antioxidant vitamin composition of pomegranate seed and juice. Int Food Res J 2015; 22: 1212-7.

[39]   Maheshu V, Priyadarsini DT, Sasikumar JM. Antioxidant capacity and amino acid analysis of

Caralluma adscendens (Roxb.) Haw var. fimbriata (wall.) Grav. & Mayur. aerial parts. J Food Sci Technol 2014; 51(10): 2415-24.
[http://dx.doi.org/10.1007/s13197-012-0761-5] [PMID: 25328180]

[40] Ambigaipalan P, de Camargo AC, Shahidi F. Identification of phenolic antioxidants and bioactives of pomegranate seeds following juice extraction using HPLC-DAD-ESI-MS^n. Food Chem 2017; 221: 1883-94.
[http://dx.doi.org/10.1016/j.foodchem.2016.10.058] [PMID: 27979177]

[41] Li R, Chen XG, Jia K, Liu ZP, Peng HY. A systematic determination of polyphenols constituents and cytotoxic ability in fruit parts of pomegranates derived from five Chinese cultivars. Springerplus 2016; 5(1): 914.
[http://dx.doi.org/10.1186/s40064-016-2639-x] [PMID: 27386358]

[42] Souleman AM, Ibrahim GE. Evaluation of Egyptian pomegranate cultivars for antioxidant activity, phenolic and flavonoid contents. Egypt Pharmaceut J 2016; 15: 143-9.
[http://dx.doi.org/10.4103/1687-4315.197582]

[43] Bhandary SK, Kumari S, Bhat VS, Sharmila KP, Bekal MP. Preliminary phytochemical screening of various extracts of *Punica granatum* peel, whole fruit and seeds. J Health Sci (Sarajevo) 2012; 2: 35-8.

[44] Manasathien J, Indrapichate K, Intarapichet KO. Antioxidant activity and bioefficacy of pomegranate *Punica granatum* Linn. peel and seed extracts. GJP 2012; 6: 131-41.

[45] Orak HH, Yagar H, Isbilir SS. Comparison of antioxidant activities of juice, peel, and seed of pomegranate (*Punica granatum* L.) and inter-relationships with total phenolic, tannin, anthocyanin, and flavonoid contents. Food Sci Biotechnol 2012; 21: 373-87.
[http://dx.doi.org/10.1007/s10068-012-0049-6]

[46] Qu W, Pan Z, Ma H. Extraction modeling and activities of antioxidants from pomegranate marc. J Food Eng 2010; 99: 16-23.
[http://dx.doi.org/10.1016/j.jfoodeng.2010.01.020]

[47] Fischer UA, Carle R, Kammerer DR. Identification and quantification of phenolic compounds from pomegranate (*Punica granatum* L.) peel, mesocarp, aril and differently produced juices by HPLC-DAD-ESI/MS(n). Food Chem 2011; 127(2): 807-21.
[http://dx.doi.org/10.1016/j.foodchem.2010.12.156] [PMID: 23140740]

[48] García-Villalba R, Espín JC, Aaby K, *et al.* Validated method for the characterization and quantification of extractable and nonextractable ellagitannins after acid hydrolysis in pomegranate fruits, juices, and extracts. J Agric Food Chem 2015; 63(29): 6555-66.
[http://dx.doi.org/10.1021/acs.jafc.5b02062] [PMID: 26158321]

[49] Mena P, Calani L, Dall'Asta C, *et al.* Rapid and comprehensive evaluation of (poly)phenolic compounds in pomegranate (*Punica granatum* L.) juice by UHPLC-MSn. Molecules 2012; 17(12): 14821-40.
[http://dx.doi.org/10.3390/molecules171214821] [PMID: 23519255]

[50] Robards K, Antolovich M. Analytical chemistry of fruit bioflavonoids. Analyst (Lond) 1997; 122: 11-34.
[http://dx.doi.org/10.1039/a606499j]

[51] Regos I, Urbanella A, Treutter D. Identification and quantification of phenolic compounds from the forage legume sainfoin (*Onobrychis viciifolia*). J Agric Food Chem 2009; 57(13): 5843-52.
[http://dx.doi.org/10.1021/jf900625r] [PMID: 19456170]

[52] Rouhi SZT, Sarker MMR, Rahmat A, Alkahtani SA, Othman F. The effect of pomegranate fresh juice *versus* pomegranate seed powder on metabolic indices, lipid profile, inflammatory biomarkers, and the histopathology of pancreatic islets of Langerhans in streptozotocin-nicotinamide induced type 2 diabetic Sprague–Dawley rats. BMC Complement Altern Med 2017; 17: 1-13.

[53] Gaber A, Hassan MM, Dessoky EDS, Attia AO. *In vitro* antimicrobial comparison of Taif and

Egyptian Pomegranate Peels and seeds extracts. J Appl Biol Biotechnol 2015; 3: 12-7.

[54] Nozohour Y, Golmohammadi R, Mirnejad R, Fartashvand M. Antibacterial activity of pomegranate (*Punica granatum* L.) seed and peel alcoholic extracts on *Staphylococcus aureus* and *Pseudomonas aeruginosa* isolated from health centers. Appl Biotechnol Rep 2018; 5: 32-6.
[http://dx.doi.org/10.29252/JABR.01.01.06]

[55] Tanveer A, Farooq U, Akram K, Shafi A, Sarfraz F, Rehman H. Antibacterial potential of pomegranate peel and seed extracts against food borne pathogens. Asian J Agri Biol 2016; 4: 60-4.

[56] Gill NS, Dhawan S, Jain A, Arora R, Bali M. Antioxidant and anti-ulcerogenic activity of wild *Punica granatum* ethanolic seed extract. Res J Med Plant 2012; 6: 47-55.
[http://dx.doi.org/10.3923/rjmp.2012.47.55]

[57] Sadeghi N, Jannat B, Oveisi MR, Hajimahmoodi M, Photovat M. Antioxidant activity of Iranian pomegranate (*Punica granatum* L.) seed extracts. J Agric Sci Technol 2009; 11: 633-8.

[58] Surveswaran S, Cai YZ, Corke H, Sun M. Systematic evaluation of natural phenolic antioxidants from 133 Indian medicinal plants. Food Chem 2007; 102: 938-53.
[http://dx.doi.org/10.1016/j.foodchem.2006.06.033]

[59] Górnaś P, Pugajeva I. Seglina. Seeds recovered from by-products of selected fruit processing as a rich source of tocochromanols: RP-HPLC/FLD and RP-UPLC-ESI/MS n study. Eur Food Res Technol 2014; 239: 519-24.
[http://dx.doi.org/10.1007/s00217-014-2247-3]

[60] Devatkal SK, Naveena BM. Effect of salt, kinnow and pomegranate fruit by-product powders on color and oxidative stability of raw ground goat meat during refrigerated storage. Meat Sci 2010; 85(2): 306-11.
[http://dx.doi.org/10.1016/j.meatsci.2010.01.019] [PMID: 20374904]

[61] Devatkal SK, Narsaiah K, Borah A. Anti-oxidant effect of extracts of kinnow rind, pomegranate rind and seed powders in cooked goat meat patties. Meat Sci 2010; 85(1): 155-9.
[http://dx.doi.org/10.1016/j.meatsci.2009.12.019] [PMID: 20374879]

[62] Narsaiah K, Jha SN, Devatkal SK, Borah A, Singh DB, Sahoo J. Tenderizing effect of blade tenderizer and pomegranate fruit products in goat meat. J Food Sci Technol 2011; 48(1): 61-8.
[http://dx.doi.org/10.1007/s13197-010-0127-9] [PMID: 23572717]

[63] Qin YY, Zhang ZH, Li L, *et al.* Antioxidant effect of pomegranate rind powder extract, pomegranate juice, and pomegranate seed powder extract as antioxidants in raw ground pork meat. Food Sci Tech 2013; 22: 1063-9.
[http://dx.doi.org/10.1007/s10068-013-0184-8]

[64] Keşkekoğlu H, Üren A. Inhibitory effects of pomegranate seed extract on the formation of heterocyclic aromatic amines in beef and chicken meatballs after cooking by four different methods. Meat Sci 2014; 96(4): 1446-51.
[http://dx.doi.org/10.1016/j.meatsci.2013.12.004] [PMID: 24398004]

[65] Kaur S, Kumar S, Bhat ZF, Kumar A. Effect of pomegranate seed powder, grape seed extract and tomato powder on the quality characteristics of chicken nuggets. NUFS 2015; 45: 583-94.
[http://dx.doi.org/10.1108/NFS-01-2015-0008]

[66] Özalp Özen B, Eren M, Pala A, Özmen İ, Soyer A. Effect of plant extracts on lipid oxidation during frozen storage of minced fish muscle. Int J Food Sci Technol 2011; 46: 724-31.
[http://dx.doi.org/10.1111/j.1365-2621.2010.02541.x]

[67] Dib A, Kasprzak K, Wójtowicz A, *et al.* The effect of pomegranate seed powder addition on radical scavenging activity determined by TLC–DPPH test and selected properties of gluten-free pasta. J Liq Chromatogr Relat Technol 2018; 41: 364-72.
[http://dx.doi.org/10.1080/10826076.2018.1449058]

[68] Ersöz E, Kınık Ö, Yerlikaya O, Açu M. Effect of phenolic compounds on characteristics of strained

yoghurts produced from sheep milk. Afr J Agric Res 2011; 6: 5351-9.

[69]   Saki AA, Rabet M, Zamani P, Yousefi A. The effects of different levels of pomegranate seed pulp with multi-enzyme on performance, egg quality and serum antioxidant in laying hens. Iran J Appl Anim Sci 2014; 4: 803-8.

[70]   Modaresi J, Fathi Nasri MH, Rashidi L, Dayani O, Kebreab E. Short communication: Effects of supplementation with pomegranate seed pulp on concentrations of conjugated linoleic acid and punicic acid in goat milk. J Dairy Sci 2011; 94(8): 4075-80.
[http://dx.doi.org/10.3168/jds.2010-4069] [PMID: 21787942]

[71]   Razzaghi A, Naserian AA, Valizadeh R, *et al.* Pomegranate seed pulp, pistachio hulls, and tomato pomace as replacement of wheat bran increased milk conjugated linoleic acid concentrations without adverse effects on ruminal fermentation and performance of Saanen dairy goats. Anim Feed Sci Technol 2015; 210: 46-55.
[http://dx.doi.org/10.1016/j.anifeedsci.2015.09.014]

CHAPTER 2

Application of Enterococci and their Bacteriocins for Meat Biopreservation

Olfa Ben Braïek[1,*], Paola Cremonesi[2] and Stefano Morandi[3]

[1]*Laboratory of Transmissible Diseases and Biologically Active Substances (LR99ES27), Faculty of Pharmacy, University of Monastir, Tunisia*

[2] *Institute of Agricultural Biology and Biotechnology, Italian National Research Council (CNR IBBA), Lodi, Italy*

[3] *Institute of Sciences of Food Production, Italian National Research Council (CNR ISPA), Milan, Italy*

Abstract: Nowadays, consumers are more aware and conscious about health concerns related to foods, which increase their demand for more safe food, particularly meats, free of additives such as preservatives, and if so with natural ones. In line with this, bacteriocinogenic lactic acid bacteria (LAB) and their bacteriocins have been widely screened and studied in the last years in view of their use in meat biopreservation. This chapter presents an emphasised overview regarding enterococci and their produced bacteriocins (enterocins) as part of interesting LAB and biomolecules with promising potentialities to be used in meat preservation as alternatives to synthetic preservatives thus satisfying consumers' demand for healthy and "bio" meat. Indeed, the characteristics of enterococci and enterococcal bacteriocins were described based on published literature. Further, we have reviewed some of the research on their applications for biopreservation of meat and meat products with a focused discussion on diverse topics such as their advantages as well as the challenges and limits of their use in meat. Finally, the synergistic approaches based on combinations of enterococcal protective cultures and/or enterococcal bacteriocins with other technological concepts to improve safety and quality of meats are reported and discussed.

Keywords: Application, Bacteriocin, Biopreservation, Biopreservative, Enterococci, *Enterococcus*, Enterocin, Lactic acid bacteria, *Listeria monocytogenes*, Meat, Pathogen, Protective culture, Spoilage.

INTRODUCTION

Microorganisms represent a risk to human health from food-borne illnesses and a problem to economic losses. Considering this, chemical additives are intensively

* **Corresponding author Ben Braïek Olfa:** Laboratory of Transmissible Diseases and Biologically Active Substances (LR99ES27), Faculty of Pharmacy, University of Monastir, Tunisia; E-mail: olfa_bbraiek@yahoo.fr

used to inhibit microbial proliferation and extend the food shelf-life [1].

As a result, several studies demonstrated that synthetic food preservatives have been linked to toxicological problems and diseases (allergic reactions, heart diseases, neurological problems and cancers) [2]. Besides, consumers are increasingly demanding safe and "bio" foods without chemical preservatives which encourage food industries to search and apply for novel strategies based on ensuring food safety and extending their shelf-life with natural antimicrobials that fall in the principle of "food biopreservation".

Several bacteria could produce antimicrobial substances called "bacteriocins", but those produced by lactic acid bacteria (LAB) have gained a great interest in recent researches [2, 3]. Among these LAB, enterococci and their produced bacteriocins called "enterocins" have received considerable attention and were extensively studied for their potentialities to be used in food biopreservation thanks to their large spectrum of antimicrobial activities against many food-borne pathogens and spoilage bacteria such as *Listeria monocytogenes*, *Pseudomonas aeruginosa*, *Escherichia coli*, *Staphylococcus aureus* and *Bacillus* spp [4].

For this purpose, the application of enterococci in foods could be realised according to two methods; (i) direct application of the bacteriocin-producing strain into food matrix as bioprotective culture or (ii) direct application of cell-free supernatant (CFS), partially purified or purified bacteriocin as a food preservative [5, 6]. These techno-biological strategies need in-depth studies regarding the safety aspects of the inoculated antimicrobials supported by toxicological data, their activity and efficacy in foods and their bactericoin production process prior to their legal approval by the authorities as applicable preservative agents. On the other hand, the incorporation of enterococci or their enterocins in packaging could be another technique to ensure microbiological quality and safety of foods [5, 6].

Among the foods that were much studied in the last years to be inoculated with enterococci and/or their enterocins to improve their overall safety, we could notice meats which represent an important source of valuable nutrients in the human diet [6, 7]. However, they are characterised by a very short shelf-life due to their composition ideal to various microbial proliferation and contamination leading to a health risk for consumers, a degradation of organoleptic quality (appearance, texture, odour, flavour, and colour) and an economic loss in meat industries.

This chapter will cover general features on enterococci and their bacteriocins as natural "solutions" attempting to solve these issues. Nevertheless, researchers and industries will always face conflicting challenges to assure meat safety and meet consumers' satisfaction. Furthermore, earlier works [8, 9] generally illustrated the role of enterococci and/or enterocins in diverse foods; however, a fewer [10] emphasised their functionalities in meat and meat products. With this aim, the present chapter summarises the most relevant insights obtained during the last years about the practical importance of enterococci and enterocins for use in biopreservation of meat as food example and highlights in detail their current applications in the meat industry by discussing both advantages and limitations and finally exposes some approaches with high hope to mainly overcome these limitations.

General Characteristics of Enterococci

Enterococci are Gram-positive, catalase and oxidase-negative, facultative anaerobic and non-spore-forming bacteria [11]. Until now, this LAB genus contains about 37 identified species, but *E. faecium* and *E. faecalis* remain the most abundant among them [12].

Enterococcus species can grow at a temperature range from 10°C to 45°C in aerobic conditions [13]. They can also grow in a wide range of pH (4.4-9.6) and tolerate media with 6.5% of NaCl and the presence of 40% (w/v) of bile salts [14].

Enterococci usually inhabit the alimentary tract of humans and animals, and can be present in various environmental sources such as water, soil and plants in addition to being isolated from different foods (dairy products, fermented vegetables, fish, seafoods, meat and meat products) [12].

Bacteriocins Produced by Enterococci

General Characteristics and Inhibitory Spectrum

Bacteriocins produced by enterococci are ribosomally synthesised peptides or proteins, cationic and amphiphilic in nature, pH and heat stable, and characterised by antimicrobial activities against closely related bacterial species [15 - 17]. However, recent studies demonstrated that these bacteriocins could have a large spectrum of action including other Gram-positive bacteria such as *L. monocytogenes* and *S. aureus* [18 - 22], Gram-negative bacteria such as *P. aeruginosa* and *E. coli* [19, 21 - 23], fungi and yeasts [21, 22] and in some cases viruses such as herpes viruses HSV-1 and HSV-2 [24 - 27], HSA virus [27] as well as polio and measles viruses [25].

Baceriocins of *Enterococcus* spp. are generally called "enterocins" but often we can note other nominations such as "mundticin" (referring to a bacteriocin that is produced by *E. mundtii*) or "durancin" (referring to a bacteriocin that is produced by *E. durans*) (Table **1**).

Table 1. Some bacteriocins produced by *Enterococcus* species.

Bacteriocin	Produced by	Enterocin classification		Isolated from	Reference
		Class	**Subclass**		
Cytolysin (CylL and CylS)	*E. faecalis*	Class I (Lantibiotic enterocins)	-	Clinical isolate	[28]
Enterocin W (Wα and Wβ),	*E. faecalis* NKR-4-1	Class I	-	Thai fermented fish	[29]
Enterocin A	*E. faecium* HTE5	Class II	II.1	Skin of gilthead bream	[30]
Enterocin P	*E. lactis* Q1	Class II	II.1	White raw shrimp (*Penaeus vannamei*)	[17]
	E. faecium P13	Class II	II.1	Dry-fermented sausage	[31]
Enterocin P and Mundticin KS	*E. mundtii* CWBI-1431	Class II	II.1	Artisanal-produced Peruvian cheeses	[32]
Enterocin CRL35	*E. faecium* CRL35	Class II	II.1	Cheese	[31]
Bacteriocin N15	*E. faecium* N15	Class II	II.1	Nuka	[31]
Bacteriocin ST5Ha	*E. faecium* ST5Ha	Class II	II.1	Smoked salmon	[26]
Mundticin ATO6	*E. mundtii* ATO6	Class II	II.1	Vegetables	[31]
Mundticin KS	*E. mundtii* NFRI 7393	Class II	II.1	Grass silage	[31]
Enterocin Q	*E. faecium* L50	Class II	II.2	Spanish dry-fermented sausage	[33]
Enterocin F-58	*E. faecium* F-58	Class II	II.2	Moroccan soft farmhouse goat's cheese « Jben »	[34]
Enterocin B	*Enterococcus faecium* T136	Class II	II.3	Spanish dry-fermented sausages	[35]
Enterocins 1071A and 1071B	*E. faecalis* BFE 1071	Class II	II.3	Faeces of minipigs	[31]

(Table 1) cont.....

Bacteriocin	Produced by	Enterocin classification		Isolated from	Reference
		Class	Subclass		
Enterocins A and B	E. faecium MMT21	Class II	II.1 and II.3 respectively	Tunisian rigouta cheese	[36]
	E. faecium P21	Class II	II.1 and II.3 respectively	Chorizo	[31]
	E. italicus GGN10	Class II	II.1 and II.3 respectively	Bovine raw milk	[14]
	E. durans GM19	Class II	II.1 and II.3 respectively	Fresh fish viscera	[37]
Enterocins A, B and P	E. faecium GM12	Class II	II.1, II.3 and II.1 respectively	Fresh fish viscera	[37]
	E. lactis 4CP3	Class II	II.1, II.3 and II.1 respectively	Pink raw shrimp (Palaemon serratus)	[22]
	E. lactis C43	Class II	II.1, II.3 and II.1 respectively	Pink raw shrimp (Palaemon serratus)	[38]
Enterocins A, B, P and X	E. faecium MMRA	Class II	II.1, II.3, II.1 and II.1 respectively	Traditional Tunisian fermented dairy product "Rayeb"	[39]
Enterocins A and B, and mundticin KS	E. faecium CWBI-B1430	Class II	II.1, II.3 and II.1 respectively	Artisanal-produced Peruvian cheeses	[32]
Enterocin L50B and Dur 152A	E. durans 152	Class II	II.2 and II.1 respectively	Floor drain sample from a food processing facility	[40]
Enterocin AS-48	E. faecalis subsp. liquefaciens S-48	Class III	-	Porcine intestinal tract	[31]
Enterolysin A	E. faecalis 2333	Class IV	-	Fish	[41]
Enterocin 012	E. gallinarum	unknown		Duodenum of ostrich	[23, 31]
Enterocin B1	E. faecium B1	unknown		Malaysian mould-fermented product tempeh	[42]
Bcateriocin BacFL31	E. faecium FL31	unknown		Fermented vegetables	[18]
Durancin GM18	E. durans GM18	unknown		Fresh fish viscera	[37]

Classification of Enterocins

In general, LAB bacteriocins were firstly classified by Klaenhammer [43] into four classes (I-IV): Class I bacteriocins (< 5 kDa) or "lantibiotics" containing non-standard amino acids, such as lanthionine and β-methyllanthionine, resulted from some post-translational modifications, with nisin A as representative. Class II bacteriocins (5-10 kDa) are heat-stable unmodified peptides. This class is divided into three subclasses: class IIa (pediocin-like bacteriocins, such as pediocin PA-1/AcH, possessing a cationic and hydrophile region with the consensus amino acid sequence of YGNGV in the N-terminal extremity known as "pediocin box motif" and a disulphide bridge formed by two cysteines in the N-terminal extremity), class IIb (two-peptide bacteriocins including lactocins 705α and 705β), and class IIc (other one-peptide bacteriocins that require a reduced cysteine for activation such as enterocin B). Class III contains thermo-sensitive proteins with molecular masses > 30 kDa. Finally, class IV consists of peptides whose activity requires a complex of molecules with lipid or carbohydrate fractions together with the protein fraction such as glycocin F produced by *Lactobacillus plantarum* KW30 [44] and enterocin F4-9 produced by *Enterococcus faecalis* F4-9 [45].

Afterward, many authors have revised this LAB bacteriocins' classification by addition or removal of some classes with slight differences in the description of subclasses resulting in different schemes based on their primary structures, molecular weights, post-translational modifications, genetic and biochemical characteristics [46 - 51]. Nevertheless, there is no universally adopted one of them.

Otherwise, it has been shown that some criteria used to classify LAB bacteriocins are not applicable to enterocins since they could have characteristics common to more than one of the previously described classes or subclasses. For instance, enterocin P could be included in subclass IIc LAB bacteriocins since it is produced by a pre-peptide translocase, but also in subclass IIa because it possesses the YGNGV consensus sequence typical of this LAB bacteriocins' subclass [9].

Hence, Franz *et al.* [52] proposed a new classification for enterocins and other bacteriocins produced by *Enterococcus* spp. taking the original and general Klaenhammer classification of LAB bacteriocins as a basis. Franz *et al.* [52] classification is the most used one for enterocins as it allows grouping the known enterococcal bacteriocins meeting thus the characteristics mainly used for previous classifications. In addition, this new enterocin classification is ascribed as the most simplified classification scheme based on enterocin structure, amino acid sequence similarities as well as their post-translational modifications [53 - 56]. Indeed, according to Franz *et al.* [52], enterocins are classified into four distinct classes named class I enterocins, class II enterocins, class III enterocins and class IV enterocins.

Enterocins belonging to class I are called "lantibiotics enterocins" since they contain lanthionine and are small peptides (< 5 kDa). The only two lantibiotic enterocins known until now are cytolysin (CylL and CylS) and enterocin W (Wα and Wβ), which are produced by *E. faecalis* strains and are formed by two peptides necessary for their action [28, 29].

Class II enterocins are nonlantibiotic peptides and are subdivided into three subclasses (Class II.1, Class II.2 and Class II.3). Class II.1 enterocins are anti-*Listeria* peptides, recognised as enterocins of the pediocin family due to their possession of the YGNGV consensus sequence. It is important to note in this context that the most abundant enterocins belong to this subclass and are considered to be of the greatest interest for use as food preservatives [9]. This class II.1 enterocins is divided into two sub-groups; sub-group 1 includes enterocins that possess an ABC transporter for their secretion and contain seven unique conserved amino acid residues such as enterocin A with structural characteristics making it equivalent to pediocin PA-1/AcH, and sub-group 2 includes enterocins that are produced via a mature pre-protein and contain six unique conserved amino acid residues such as enterocin P [52]. Class II.2 enterocins did not possess the YGNGV consensus sequence nor the ABC transporter as secretion system and are synthesised without a leader peptide. This enterocins' subclass II.2 is further divided into two sub-groups; sub-group 1 including monomeric proteins such as enterocin Q and sub-group 2 including enterocins that need the formation of an heterodimeric active complex such as enterocin L50 [52]. Class II.3 enterocins groups the other linear and non-pediocin type enterocins that are synthesised with leader peptide such as enterocin B and enterocins 1071A and 1071B [52].

Enterocins that belong to class III are cyclic antimicrobial peptides like enterocin AS-48 [52]. Finally, class IV enterocins includes heat-labile proteins with high molecular weight (34.5 kDa) such as enterolysin A produced by *E. faecalis* 2333 [52, 57].

As mentioned above, it could be noticed that well characterised enterocins (*e.g.* enterocins A and B) could be easily grouped in the Klaenhammer classification of LAB bacteriocins, but some others (*e.g.* enterocin AS-48, enterocins L50A and L50B) with structure and genetic traits different from known bacteriocins produced by other LAB, could not be grouped into this classification scheme. That's why the enterocin classification suggested by Franz *et al.* [52] seems to be the most suitable to well describe all enterococcal bacteriocins into such a classification system. However, when comparing enterocins with other LAB bacteriocins belonging to "similar" classes (based on Klaenhammer [43] and Franz *et al.* [52] classifications), some differences could be detected. For example, cytolysin and enterocin W belonging to class I enterocins (lantibiotic enterocins) as they contain lanthionine, are two-component bacteriocins consisting of two linear peptides that structurally differ from nisin A, a lantibiotic bacteriocin which consists of only one linear peptide. In addition, although enterocin P and pediocin PA-1/AcH are recognised as pediocin-like or *Listeria*-active bacteriocins with YGNGV consensus sequence, the first enterococcoal bacteriocin (enterocin P) possesses an ABC-type transporter protein implicated in its export, while pediocin PA-1/AcH is exported by a pre-protein translocase.

Interestingly, it is important to mention that enterococci have the capacity to produce multiple bacteriocins simultaneously (Table **1**). The well-known enterocins' co-productions are observed with enterocins A and B, and enterocins A, B and P [14, 22, 31, 32, 37 - 39].

Mechanism of Action

Several mechanisms have been described for the bactericidal mode of action of enterocins which differ from one class to another involving, in general, the presence of molecular receptors in the membrane of the susceptible cell. Effectively, the cytoplasmic membrane represents the primary target of enterocins, as most bacteriocins [58]. However, it is important to denote that the mode of action of those belonging to class II.1. enterocins (enterocins of the pediocin family) has been thoroughly studied.

Class I enterocins (lantibiotic enterocins), such as cytoloysin, have the pore-forming mode of action, similar to lacticin 3147, the well-studied LAB lantibiotic bacteriocin produced by *Lactococcus lactis* DPC3147 [59]. This mode of action implicates cytolysin binding to lipid II, a universal receptor, by its amino terminals resulting in a complex able to initiate the process of membrane insertion and formation of pores leading to rapid cell death [60]. Whereas, the cytolysin mode of action varies from that of nisin (class I LAB bacteriocin) which binds to lipid II and inhibits peptidoglycan synthesis, the main component of cell wall thus causing cell death [9].

Class II enterocins (nonlantibiotic enterocins) are known by their amphiphilic structure that allows them to be inserted into the cytoplasmic membrane of the target cell by its permeabilisation by the formation of pores which disrupts the integrity of the cell leading to the out leakage of K^+ ions and other low-molecular weight molecules, dissipation of trans-membrane potential, depletion of pH gradient and intracellular ATP and inhibition of amino acid uptake thus resulting in cell death [12, 9, 61, 62]. This membrane-permealising mode of action has been observed in enterocin P [31, 63] and bacteriocin ST5Ha [26].

Studies on the mode of action of class III enterocins (cyclic enterocins) have been intensively performed on enterocin AS-48 and showed that its mechanism of antimicrobial action is based on a direct electrostatical interaction with the target cell membrane without the need of any receptor molecule, causing its permeation known as electropration by the formation of non-selective channels in the cytoplasmic membrane leading to the leakage of ions, depolarisation and eventually cell death [64 - 67]. Since enterocin AS-48 is the only cyclic enterocin belonging to class III enterocins currently studied for its mode of action, so it is not possible to generalise its antimicrobial mechanism to all class III enterocins. Otherwise, when comparing the current mode of action of enterocin AS-48 with those of other cyclic LAB bacteriocins, some similarities and differences could be ascribed. Effectively, gassericin A, produced by *Lactobacillus gasseri*, induces cell death by the leakage of K^+ ions resulting from the permeation of target cell membrane, similarly to enterocin AS-48 [68]. However, it has been shown that carnocyclin A, produced by *Carnobacterium maltaromaticum* UAL307, forms anion-selective and voltage-dependent pores on the membranes of sensitive cells [69]. Generally, it is true that the modes of action of most cyclic bacteriocins have not yet been studied, but until now, it could be concluded that these three cyclic bacteriocins, do not require membrane receptors for their bactericidal action against target cells [64, 68, 69].

Finally, the mode of action of class IV enterocins, such as enterolysin A, is quite different from the other enterocins because it attacks susceptible bacteria by degrading the cell wall structure with hydrolysis of peptide bonds of peptidoglycan, which eventually leads to lysis of the cells of target strains [41].

Application of *Enterococci* and their Bacteriocins in Meat

Challenges of their Application in Meat

Nowadays, the application of synthetic preservatives in foods remains a delicate issue because of their implication in toxicological and carcinogenic problems [1, 2]. Consequently, food industries are currently investigating new solutions as alternatives to chemicals since consumers become more conscious about health concerns and are demanding safe foods without artificial additives or with natural additives. This has driven food companies to explore natural antimicrobials for "bio", "green" or "clean" labels [70]. Considering this, the application of antimicrobial peptides or proteins gained great attention as food preservatives particularly in meat and meat products [6]. Effectively, only nisin A (Nisaplin®), pediocin PA-1/AcH (ALTA™ 2341), lysozyme and lactoferrin are allowed to be used as natural additives in meats [71, 72]. In fact, strict *in vitro* and *in vivo* tests must be realised rigorously prior to the final decision of the efficacy of the new natural antimicrobial in meat matrices. These tests could take many years and require expensive costs before they could be completed to finally obtain approval for the new food biopreservatives from the international regulation agencies [70].

Advantages of their Application in Meat

In these recent years, the application of lactic acid bacteria (LAB) and/or their produced bacteriocins in biopreservation of foods (particularly meats) is a promising biotechnological approach to improve their safety and maintain their shelf-life as they are *"Generally Recognised as Safe"* microorganisms [5, 73]. Deep researches on LAB potentialities in the preservation of various meat systems led to discoveries of interesting results for *enterococci* and their produced antimicrobial peptides suggesting their application in meat industries [18, 74, 75]. Indeed, the direct application of enterocin-producing *Enterococcus* strains as protective cultures in meat is a very practical and economical trend from regulatory status. In fact, this advanced approach did not require several processing steps (*e.g.*, production and purification) nor many legal restrictions compared to the direct application of their semi-purified or purified bacteriocins [74].

On the other hand, in contrast to other proteins like lysozyme and other LAB bacteriocins such as nisin A and pediocin PA-1/AcH used as preservatives, enterococcal bacteriocins have several advantages for use in meat biopreservation. In this sense, it is known that enterocins are more stable in extreme levels of environmental conditions. Indeed, they are the most heat resistant among the other LAB bacteriocins, and could tolerate high thermal treatments without being denaturated or losing their antimicrobial activity [6]. In addition, enterocins remain active at low temperatures for extended periods of time, at a wide range of pH and at very low concentrations (picomolar and nanomolar) even when directly added to meat matrix [9, 76]. Furthermore, the advantageous use of enterocins in meats is the fact that they did not modify the organoleptic quality of meat since they are odourless, colourless and tasteless peptides. Other merits played by enterocins as most interesting bacteriocins for meat preservation rather than other LAB bacteriocins, could be also noticed. In fact, the great distribution and natural emergence of enterocin-producing strains in many food systems and environmental sources as well as in human and animal gastrointestinal tracts, as commensal microorganisms, make their isolation and screening for enterocin production much easier [6]. Moreover, until now, many of the characterised enterocins have shown a large spectrum of action and strong bactericidal activity against undesirable and pathogenic microorganisms present in foods such as *L. monocytogenes*, *S. aureus*, *Clostridium* sp., *E. coli*, *Bacillus cereus* suggesting that enterocins could have important functions as natural preservatives in meat

[9]. Finally, these enterococcal bacteriocins are known to be safe for human consumption as they are not toxic, could be digested by proteases and naturally occur in the microbiota of various foods [70]. Taken into account all of these potent traits, *enterococci* and their produced enterocins are attracting considerable interest to be used as innovative natural food preservatives to ensure safety and extend shelf-life of meats.

Current Applications of Enterococci and their Enterocins in Meat Preservation

Direct Application of Enterocin-producing Enterococci in Meat

Meat is one of the most perishable foods [77]. Its bacterial spoilage is influenced by extrinsic parameters related to environmental and storage conditions like temperature, water activity and oxygen availability, and by intrinsic parameters such as its nutrient composition, pH, moisture content and the presence of microorganisms in terms of quantity and types [7]. All of these factors result in the development of undesirable odours and flavours, dark colour, formation of slime and production of gas making meats unappetising and unacceptable for human consumption.

Spoilage bacteria that could occur in meats consist of *Brochothrix thermosphacta*, *Carnobacterium* spp., *Weissella* spp., *Streptococcus* spp. and rarely *Bacillus* spp [78, 79]. In general, these spoilage microbes do not harmfully affect human health but when consumed in high concentrations they could cause gastrointestinal disturbances [79]. Yeasts and moulds are also among the spoilage microorganisms that commonly infect meats such as *Penicillium*, *Mucor*, *Cladosporium* and *Alternaria* [78].

Otherwise, it is important to notice in this meat spoilage context that some chemical reactions catalysed by oxygen, salt, temperature and oxygen reactive species could occur during the storage or processing of meat [80]. The most known reactions that naturally take place in fresh and cooked types of meat are oxidation of lipids and proteins present in the meat which finally end up in meat decomposition resulting in harmful chemical compounds (*e.g.*, hydroperoxides, carbonyls, hydrocarbons, furans, oxidized proteins, metmyoglobin, nitrosamines, heterocyclic amines, biogenic amine, *etc.*) [81]. These harmful chemicals lead to the meat quality deterioration and induce significant toxicological effects to consumers that range in severity from uncomplicated effects such as nausea, migraine and rashes to more dangerous ones such as respiratory distress, atherosclerosis, neurodegenerative diseases and cancer [81, 82].

Pathogenic bacteria generally infecting meats when processed, cut, packaged, transported, stored, sold and/or handled include Gram-positive bacteria such as *L.*

monocytogenes, S. aureus, Clostridium botulinum, Clostridium perfringens, Bacillus cereus, and Gram-negative bacteria like *P. aeruginosa, E. coli, Yersinia enterocolitica, Campylobacter jejuni, Salmonella* spp., *Shigella* spp., *Acinetobacter* spp., *Enterobacter* spp., *Moraxella* spp. and *Proteus* spp., causing severe health problems and diseases that in some cases can lead to death [5, 79]. Hence, their implication in diverse foodborne outbreaks worldwide has raised consumer questions on the safety of meats.

Even though there are many enterocin-producing enterococci isolated from different origins that marked a great interest amongst researchers and biotechnologists, only a few were studied at a pilot-scale in terms of laboratory challenge tests into meat samples to evaluate their efficacy on meat biopreservation (Table **2**).

As protective cultures, the enterocin-producing *Enterococcus* spp. aimed to inhibit the proliferation of diverse spoilage and pathogenic microorganisms without altering the sensorial quality of meats.

Previously, it was reported by Callewaert *et al.* [83] that *E. faecium* RZS C13 and *E. feacium* CCM 4231 inhibit the growth of *Listeria* spp. in dry-fermented sausages. Also, these two enterococci improved the sensorial properties of the tested sausages suggesting thus their application as protective and starter-cultures in dry-fermented sausages.

E. faecium CTC 492 producing enterocins A and B was assessed as a biopreservative culture on the prevention of the slime production in sliced vacuum-packed cooked pork [84]. As a result, it was found to be efficient in preventing ropiness due to *Lactobacillus sakei* CTC 746 in sliced cooked ham until 21 days of storage at 8°C. This study showed that *in situ* produced enterocins A and B performed better than nisin and sakacin K and demonstrated no inhibitory effect on the production of slime.

The bacteriocinogenic *E. casseliflavus* IM 416K1 isolated from Italian sausages was used in diverse trials in order to evaluate its effect on the growth of *L. monocytogenes* NCTC 10888 in artificially inoculated Italian sausages "Cacciatore" [85]. Thus, experiments demonstrated that the tested enterococcal strain IM 416K1, enterocin 416K1 producer, caused a significant decrease of *L. monocytogenes* populations as compared to the control only inoculated with the pathogen.

The study of Sparo *et al.* [86] demonstrated the efficacy of *E. faecalis* CECT7121 strain to inhibit the growth of *Escherichia coli* O157:H7, *S. aureus, L. monocytogenes* and *C. perfringens* in ground beef meat samples to undetectable levels, thus achieving a significant bactericidal effect against these pathogens.

Table 2. Application of some *Enterococcus* spp. strains and/or enterococcal bacteriocins in meats.

Enterococcus spp.	Bacteriocin	Application	Reference
Application of enterococci as protective cultures			
E. faecium CCM 4231	Enterocin CCM 4231	Spanish-style dry-fermented sausages	[83]
E. faecium RZS C13	Enterocin RZS C13	Spanish-style dry-fermented sausages	[83]
E. faecium CTC 492	Enterocins A and B	Cooked pork	[84]
E. casseliflavus IM 416K1	Enterocin 416K1	Italian sausages (Cacciatore)	[85]
E. faecalis CECT7121	Enterocin AP-CECT7121	Ground beef meat	[86]
E. lactis 4CP3	Enterocins A, B and P	Raw beef meat	[74]
Application of enterococcal bacteriocins as preservatives			
E. faecium CCM 4231	Enterocin CCM 4231	Dry-fermented Hornád salami	[87]
E. faecium CTC 492	Enterocins A and B	Dry-fermented sausages	[88]
E. faecium CTC 492	Enterocins A and B	meat products (cooked ham, minced pork, fermented sausages and pâté)	[89]
E. faecium CRL35	Enterocin CRL35	Meat system	[90]
E. faecalis A-48-32	Enterocin AS-48	Fermented sausages "fuet"	[91]
E. faecalis A-48-32	Enterocin AS-48	Cooked ham	[92]
E. faecium MT 104 and MT 162	Bacteriocins MT 104 and MT 162	Chilled sausages	[93]
E. faecium FL31	Bacteriocin BacFL31	Minced beef	[18, 75]
E. durans 152	Dur 152A and enterocin L50B	Deli ham	[40]

Recently, *E. lactis* 4CP3, a multiple enterocin-producing strain (enterocins A, B and P), was observed to strongly inhibit the growth of *L. monocytogenes* EGDe 107776 in raw beef meat during 28 days of refrigerated storage [74]. Effectively, the application of *E. lactis* 4CP3 culture at 10^7 CFU/ml significantly reduced the numbers of *L. monocytogenes* artificially contaminated in raw beef meat by 6.77 log units (from approximately 10^8 CFU/g to approximately 10^2 CFU/g) within 7 days of storage. Then, the listerial growth was completely inhibited from day 14 to the end of the challenge test. Hence, according to the authors, *E. lactis* 4CP3 could be used as natural and potential protective culture against *L. monocytogenes* growth in meat products.

Finally, it is important to note that according to the literature, the ability of enterococci to inhibit *Listeria* spp. is well known and many studies were conducted in view of proving the anti-*Listeria* activity of *Enterococcus* spp. when added in meat models thanks to their produced bacteriocins. The availability of large data may be explained by the feasibility of enterococcal tests towards *Listeria* since these microorganisms have a close phylogenetic relationship [15]. On the other hand, *L. monocytogenes* represents one of the main concerns in the meat industry. In fact, its high prevalence in meat and meat products has been intensively reported [94, 95]. This bacterium has been implicated in several listeriosis outbreaks worldwide due to the consumption of contaminated meats making it a significant foodborne pathogen and an important public health issue [96].

Direct Applications of Enterococcal Bacteriocins in Meat

The direct application of enterococcal bacteriocins is another promising approach used in the biocontrol of the growth of spoilage and pathogenic microorganisms in meats. This kind of biopreservative method is used when *in situ* antimicrobial effect of bacteriocinogenic *Enterococcus* spp. cultures was weak or lost in meat model comparing to *in vitro* results observed in culture media. In fact, in some cases, antagonistic activities of enterococci could be more apparent in microbiological media (broths and agars) than in real meat systems and enterococcal strains could not produce bacteriocins in the tested meat matrix due to diverse conditions (concentration of the enterocin-producing strain inoculum, initial concentration of intrinsic flora of the meat, interactions with meat components, storage temperature...). The forms of bacteriocin preparations that could be applied in meats consist of cell-free supernatant, partially-purified bacteriocin and purified bacteriocin [5, 6].

In this context, the enterocin CCM 4231 produced by *E. faecium* CCM 4231 was added to dry-fermented Hornád salami for the control of *L. monocytogenes* [87]. Results indicated an immediate reduction of the listerial cells. Drastic decrease continued to be observed after three weeks of ripening.

Furthermore, semi-purified enterocins A and B produced by *E. faecium* CTC 492 were applied in different meat products (cooked ham, minced pork, fermented sausages and pâté) [88]. These enterocins decreased significantly the growth of *L. innocua* during refrigerated storage. However, the addition of the enterocin-producing *E. faecium* CTC 492 as protective culture in dry-fermented sausages had no anti-listerial effect due to the growth interference from product ingredients [89].

Similarly, Vignolo *et al.* [90] reported the efficiency of enterocin CRL35 produced by *E. faecium* CRL35 in a meat system against *L. monocytogenes* and *L. innocua*.

Also, the partially purified enterocin AS-48 produced by *E. faecalis* A-48-32 was applied in fermented sausages "fuet" for the biocontrol of the pathogens growth during ripening [91]. Significant inhibitory effects on *L. monocytogenes* and *Salmonella* were observed after 10 days of ripening. Equally, enterocin AS-48 notably reduced the growth of *S. aureus* in sausages and cooked ham samples [92, 97]. Thus, all of these results indicate that enterocin AS-48 could be considered as biopreservative agent in various meat products.

On the other hand, bacteriocins MT 104 and MT 162 produced by *E. faecium* designated as MT 104 and MT 162 were reported to strongly inhibit the proliferation of *L. monocytogenes* in meat sausages after 5 days of chilled storage at 4°C [93].

In a more recent study, the application of 400 AU/g of pre-purified bacteriocin BacFL31 produced by *E. faecium* FL31 was noted to potentially prevent the contamination and the growth of *L. monocytogenes* ATCC 19117 during the storage of minced beef meat at 4°C [18].

Another recent research reported the anti-*L. monocytogenes* effect of the addition of two (each one apart) partially purified bacteriocins produced by *E. durans* 152, Dur 152A (an enterocin L50A derivative with two amino acid substitutions) and enterocin L50B, at 400 AU/ml in deli ham [40]. Indeed, results indicated that these two enterococcal bacteriocins inhibited listerial growth in deli ham for 10 weeks at 8°C and for 30 days at 15°C comparing to nisin which prevented *L. monocytogenes* growth for up to 6 weeks at 8°C and up to 18 days at 15°C. Hence, these findings revealed that the two tested bacteriocins (Dur 152A and enterocin L50B) have considerable potential to be used as biopreservatives in deli meat.

Limits of Enterococci and Enterocins Use in Meat Biopreservation

Limits of Enterococci

The main objective of bacteriocin-producing *Enterocccus* spp. strains used as protective cultures in meat is to inhibit the growth of undesirable bacteria without the cause of sensorial changes in meat products. However, in some cases, inoculated cells could produce acids, in addition to bacteriocins, and hydrolyse meat proteins which dramatically change the organoleptic characteristics (texture, appearance, odour, flavour and colour) of meat samples [6].

Another limit that could be mentioned here is the incompatibility between the protective cultures of bacteriocin-producing enterococcal strains and other inoculated cultures in meats such as starter cultures or co-cultures. That's why this aspect should be carefully and perfectly evaluated before such applications.

On the other hand, it is important to note that the lack of protective effect of *Enterococcus* spp. strains added in meat systems due to interference of the tested food matrix and the un-optimised conditions for the bacteriocin production represent a further concern associated with the application of bacteriocinogenic enterococcal producers as preservative inoculums in meat.

Finally, despite the interesting benefits that could procure enterococci as protective cultures in meat and meat products as reported in many previous studies, their implication in human diseases have generated serious concern regarding their use in food biopreservation. Indeed, nowadays, enterococci are known to be among the most common opportunistic pathogens of humans able to cause diverse nosocomial infections such as urinary infections, bacteraemia and endocarditis [98, 99]. Traits associated with their pathogenicity are virulence factors and the emergence of antibiotic-resistant enterococcal strains particularly vancomycin-resistant enterococci (VRE) [12]. In addition, since enterococci are known for their genome plasticity [100], the fact that virulence and antibiotic resistance genes could be transferred to safe enterococcal strains (free of these determinants) raises an enormous fear regarding their safety for food preservation uses which could in part explain the few numbers of commercial enterococcal strains used as feed additives [12].

Furthermore, enterococci were described to have the ability to produce biogenic amines (BA), due to the decarboxylation of amino acids, such as tyramine, histamine, putrescine and cadaverine which are unwanted compounds suspected to be toxic for consumers after an excessive oral intake causing nausea, headaches, migraine, rashes, change in blood pressure, respiratory distress, tachycardia, etc… [82, 101]. Food products that mostly contain high amounts of biogenic amines and could thus cause food poisoning are fish and cheeses [82, 102, 103]. In this context, it is important to denote that fermentation promotes the production of biogenic amines because of the acidification and proteolytic activities [104]. Indeed, cheese has received much attention since enterococci are extensively used to intensify the flavour of many kinds of cheese [105 - 107]. Equally, *Enterococcus* spp. are also applied in the production of fermented meats such as dry-fermented sausages thanks to their technological properties *e.g.* the development of characteristic aroma and flavour and the inhibition of bacterial growth [108]. Hence, this worry contributes too to the controversial status of enterococci between beneficial as protective cultures in meats or foe microorganisms?

For these reasons, enterococcal strains in view of future use as protective cultures in meat (and/or other food matrices) must be well assessed for their efficacy in meat systems by several activity validation techniques and well examined regarding their safety aspects (haemolysis, gelatinase activity, virulence factors, antiobiotic resistance, biogenic amine production, toxicological effect...) to accurately distinguish between pathogenic and safe strains. With these measures, meat industrials and consumers could be convinced by the potentiality of enterococci as promising candidates in meat biopreservation, like other LAB strains.

Limits of Enterocins

Taking into account the limits of using bacteriocinogenic enterococci in meat biopreservation, particularly their safety concerns, the use of bacteriocins (enterocins) may be considered more suitable for such an application. However, these biomolecules exhibit a number of issues and restrictions regarding their use as meat preservatives. Firstly, until now, despite the increasing numbers of new potent enterocins well characterised for their efficiency as natural protective antimicrobials, only the two bacteriocins of nisin A (Nisaplin®) and pediocin PA-1/AcH (ALTA™ 2341) have been approved by the FDA (Food and Drug Administration) and are commercially used as food preservatives in meat and meat products [72]. This could be mainly explained by the strict requirements and safety legislations taken from regulatory agencies prior to final approval of possible use of a bacteriocin in foods. This approval process is very complex including the study of acceptable daily intake, absence of adverse effects level in terms of toxicological and allergenic aspects when consumed, high antimicrobial and preservative effects when incorporated into meat systems, absence of influence on the meat sensory characteristics and absence of meat component interactions [70]. Unfortunately, collecting these informative and important data is time-consuming and is very expensive which limits the use of new bacteriocins like enterocins in meat biopreservation.

Furthermore, the effective antimicrobial functionality of such bacteriocins (enterocins) in meats could be achieved with highly purified substances (rather than semi-purified ones) and their application at large scale in terms of industrial processing will involve many steps (*e.g.* purification, production) thus increasing the cost of both of the desirable bacteriocin, although its yield is low, and the finished food product (meat and/ meat product).

Another limitation of the use of enterococcal bacteriocins in meat products is related to intrinsic factors of the food matrix (non homogeneous, high content of protein and fat, pH relatively high) that could unfavourably affect the efficacy of the tested bacteriocin [70]. Added to that, the emergence of bacteriocin-resistant microorganisms in meat samples supplemented with bacteriocins is among the limits of the application of enterocins as meat preservatives [5].

Consequently, to enhance the efficacy of added bacteriocinogenic preparations at their different forms (enterococcal cultures or enterococcal semi-purified or purified bacteriocins) in order to maintain the quality, guarantee the safety and extend the shelf-life of meat products, their application in combination with other antimicrobials (bacteriocinogenic LAB, bacteriocins, proteins, plant extracts, essential oils, nanoparticles…) or several preservation methods (low temperature, low pH, low water activity, pressure, heat, packaging) is strongly recommended. These innovative solutions are well discussed in the next section.

Combinations between Bacteriocinogenic Enterococci and/or Enterococcal Bacteriocins with Other Antimicrobials or Preservation Treatments

The suggested combinations mentioned above between enterococci and/or enterococcal bacteriocins with other control measures (physical, physicochemical, chemical or biological treatments) may result in potent synergistic activities thus being promising technological approaches that could benefit the meat industries at cost level.

Combinations between bacteriocinogenic enterococci with other bacteriocin producer LAB strains as protective cultures in meat should be realised after assessment of the compatibility between mixed strains and a realistic confirmation of their synergistic activity in meat systems.

In some cases, the addition of a producer strain as the sole protective culture in meat models does not control the growth of spoilage and pathogenic bacteria. Indeed, Aymerich *et al.* [88] observed that *E. faecium* CTC 492 added in dry-fermented sausages had no effect on the inhibition of *Listeria* growth. However, the combination of both *E. faecium* CTC 492 and its produced enterocins (A and B) significantly reduced *Listeria* counts in dry-fermented sausages and prevented ropiness in sliced cooked ham [84, 88].

Moreover, the use of a mixture of multiple bacteriocins belonging to different classes and exhibiting different mechanisms of actions is helpful not only to improve antimicrobial activity but also to avoid bacteriocin-resistances from several microorganisms. In this context, the combination between enterocin CRL35, lactocin 705 and nisin indicated a strong anti-*L. monocytogenes* FBUNT effect in the meat system showing no viable listerial counts after 3 h of incubation [90].

Regarding applications of enterocins combined with high hydrostatic pressure (HHP) or heat treatment or irradiation, several studies have been conducted in this respect. In fact, Garriga *et al.* [109] proposed an effective technological concept against the proliferation of *L. monocytogenes* in a model meat system (cooked ham) consisting of the combination between enterocins (A and B) and HHP treatment at 400 MPa for 10 min at 17°C.

Then, the study of Jofré *et al.* [110] showed that the application of enterocins A and B semi-purified from *E. faecium* CTC492 in combination with HHP at 600 MPa reduced effectively *S. enterica* and *L. monocytogenes* in both cooked and dry-cured ham.

Marcos *et al.* [111] indicated that treating sliced cooked ham with a combination of HHP (400 MPa for 10 min) and enterocins A and B during refrigerated storage decreased significantly *L. monocytogenes* viable cells.

On the other hand, it has been reported that the addition of semi-purified preparations of enterocins A and B (2000 AU/g) to raw sausages artificially contaminated with *Salmonella* combined with the application of an HHP treatment at the end of ripening considerably inhibited the *Salmonella* growth [112].

Similarly, enterocin AS-48 combined with a HHP at 400 MPa has been described to cause an important reduction of *Listeria* and *Salmonella* counts in fuet sausages during ripening and afford protection against their regrowth during storage [91]. Another finding regarding this same cyclic enterocin (AS-48) indicated its efficient synergetic effect when applied in combination with physicochemical or heat treatments on *S. aureus* and *Listeria* growth in a cooked ham model system [92].

Moreover, a strong anti-*Listeria* effect in meat sausage was observed for the enterococcal bacteriocin MT 104 with γ-irradiation [93]. In fact, it has been shown recently that irradiation combined with bacteriocin microencapsulation and/or other food preservatives were among the advanced processes to improve the food safety of meat products such as ready-to-eat (RTE) meats [6].

In a more recent study, it has been demonstrated that the semi-purified enterococcal bacteriocin BacFL31 combined with the aqueous peel onion (*Allium cepa*) extract (APOE) applied to ground beef meat was efficient against microbial proliferation, decreased the thiobarbituric acid reactive substances (TBARS), the metmyoglobin (MetMb) and carbonyl group values and delayed the disappearance of sulfhydryl proteins. The BacFL31-APOE combination contributed also to the decrease of protein as well as primary and secondary lipid oxidations, and enhanced significantly the sensory properties of treated ground beef meat [75].

Finally, it is worth mentioning that bioactive packaging containing bacteriocins have been developed as advanced technologies responding to many issues. First, they were reported to overcome the loss of bacteriocin activity when directly added into meat due to the dilution effect or because of its inactivation by some meat components such as lipids and enzymes [6, 113]. Then, these techniques are based on the release of bacteriocins from packaging to the meat matrix which avoid their direct contact with the meat product, require low incorporated amounts of bacteriocins, thus reducing their use from meat industries and responding to consumers demand for meat without or with few preservatives. Hence, the use of packaging systems into meats become helpful strategies that make the bacteriocins' applicability easier, cheaper and more potent to preserve the overall safety and quality of meat and meat products. There are two methods regarding the incorporation of bacteriocins into meat packaging systems: (1) direct bacteriocins incorporation into film matrix (cast or heat-pressed films) and (2) indirect incorporation by coating bacteriocins onto the film surface [6]. For instance, enterocins were previously incorporated into alginate, zein and polyvinyl alcohol-based biodegradable film thus increasing the safety of sliced ham by delaying or reducing the growth of *L. monocytogenes* [114].

Occasionally, we can encounter an incompatibility between bacteriocins and biopolymers in the direct incorporation of bacteriocins. For this reason, their indirect incorporation by coating was proposed. In fact, Iseppi *et al.* [115] demonstrated a marked decrease of *L. monocytogenes* NCTC 10888 viable counts in frankfurters packed with enterocin 416K1-activated coated low-density polyethylene (LDPE) films in the first 24 h of refrigerated storage. Furthermore, this kind of meat packaging seems to be beneficial since coated films are similar to uncoated ones in terms of transparency and bacteriocins coating showed good adhesion to used films even without any preliminary treatment [6].

Thus, on the whole, all of these bioactive packaging systems are biodegradable making them environmentally friendly methods, along with their low cost and feasibility in the application which prove their efficacy to be used in the meat packaging field to maintain microbiological and/or sensorial qualities and safety of meats.

Otherwise, another promising avenue which offers potential hope in overcoming some limitations of enterocins in meat biopreservation and/or other technological hurdles related to their incorporation into packaging for example, as highlighted in the preceding sections, is the use of nanotechnology. Indeed, nanoformulations prepared with interaction between bacteriocins and nanoparticles (NPs) are recently well examined and several bacteriocin-nanoconjugates have emerged as a revolutionary preservative solution in the food industry. Effectively, these bacteriocin-nanoconjugates are made by different approaches in order to improve the antimicrobial and antifungal activities of bacteriocins (*e.g.* enterocins), enhance their spectrum of action against diverse food-spoiling and pathogenic microorganisms, reduce the requirement of high bacteriocin dosage, ensure safer antimicrobial packaging and extend the shelf-life of food [116, 117]. Indeed, in food preservation, the combination of bacteriocins with nanoparticles is realised by the formation of nano-encapsulations to be protected from degradation by gastrointestinal enzymes and to improve their stability and commercial yield. The common nanotechnological approaches used to conjugate bacteriocins with nanoparticles are: (i) encapsulation of bacteriocins in nanoliposomes, (ii) conjugation of bacteriocins with chitosan nanparticules and (iii) conjugation of bacteriocins with metallic nanoparticles such as silver and gold [116, 118]. Till date, numerous LAB bacteriocins (nisin A, nisin Z, bacteriocins produced by *L. plantarum* ATM11) were subjected to these interactions types, however, only one study [119] on conjugation of enterocin FH99, produced by *E. faecium* FH99, with silver nanoparticles (En-SNPs) is currently published reporting that the enterocin-nanoconjugate showed an enhanced antimicrobial activity with a broad spectrum of action against a number of food-spoiling and pathogenic bacteria (Gram-positive and Gram-negative) without any noticeable toxicity to red blood cells.

Therefore, in this context, it could be shown that the integration of biotechnology and nanotechnology resulting in beneficial bacteriocin-nanoconjugate compounds represents a new area to potentially ensure food preservation (*e.g.* meat preservation) thus playing an important role in food safety. Nevertheless, more studies are needed to be performed, in the future, on elaborating new enterocin-nanoconjugates, optimising their formulations and examining their toxicological effects to get in-depth sufficient details about their efficiency and safety before their final approval as innovative antimicrobial compounds in food (meats and meat products) preservation which will undoubtedly increase their utility at the industry-scale level.

The current applications of *Enterococcus* spp. strains and their bacteriocins as promising protective cultures or preservatives in meat, advantages and limitations of their use and promising solutions in overcoming their limits are summarised in Fig. (**1**).

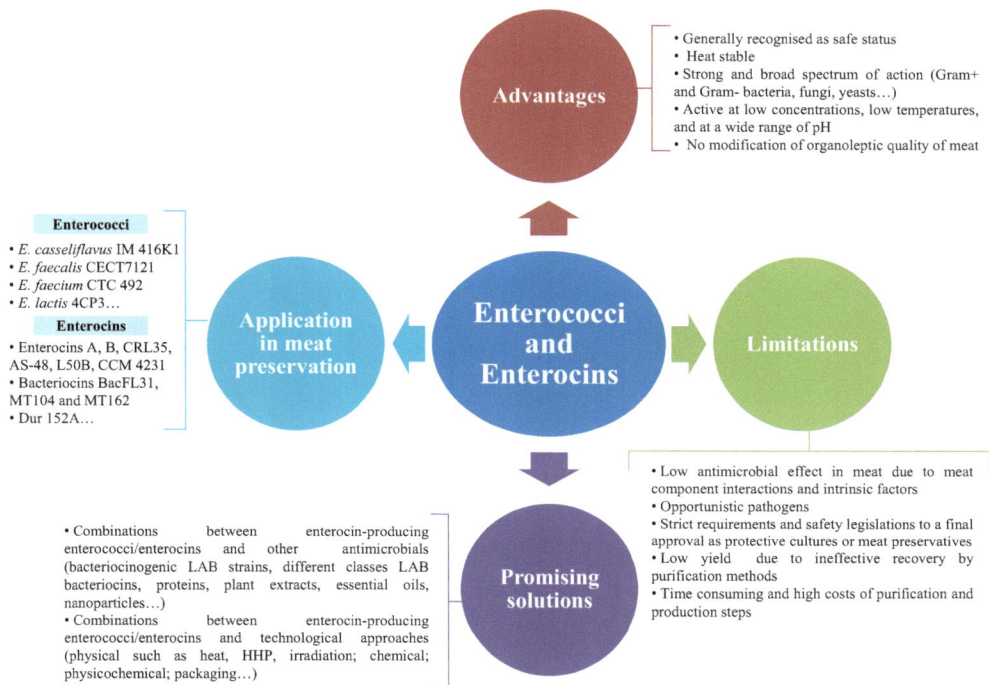

Fig. (1). Summary of the main applications of enterococci and their enterocins, their advantages, limitations and promising solutions to overcome limits. LAB: lactic acid bacteria; HHP: high hydrostatic pressure; Gram+: Gram-positive bacteria; Gram -: Gram-negative bacteria.

CONCLUSION

This chapter shows the interesting attributes of enterococci and their produced bacteriocins (enterocins) based on several research studies and published reviews making them innovative future bioagents in meat preservation alternatives to chemical preservatives. However, such natural antimicrobials will not likely serve as magic solutions to eradicate spoilage and pathogenic microorganisms in meats, because "natural" does not automatically mean "safe". Accordingly, many characterisation tests regarding enterococcal safety aspects should be thoroughly realised to confirm their innocuity and possibility to be added into meat for consumption, especially because of the serious worry disseminated between consumers about the pathogenicity of enterococci. Further experiments should be deeply carried out prior to the approval of enterococcal protective cultures and/or enterococcal bacteriocins as promising agents in meat biopreservation. These assessments include evaluation of the protective action efficacy (high bacteriocin production capacity, strong and broad antimicrobial activities in different meat models against several undesirable microorganisms), the nonalteration of meat sensorial properties and the extension of shelf-life to properly select good candidates, and much more importantly, well adapted to each particular type of meat matrix in which they will be applied.

This chapter has compiled important findings according to the literature demonstrating that many discovered strains of enterococci and their enterocins were widely studied for their effectiveness in meat biopreservation. Nevertheless, these results are still exploitable only at a laboratory scale; their functionality at industrial, meat processing and commercial levels remains unknown and strongly needs to be examined urgently. In fact, until now, only the two LAB bacteriocins of nisin A and pediocin PA-1/AcH were approved to be used as biopreservatives in meat and meat products. However, these mentioned bacteriocins exhibit some limitations that are mainly related to their low solubility in meats, their possibility of denaturation under several conditions and their narrow antimicrobial spectrum of action towards several spoilage and pathogenic microorganisms. Consequently, there is a great need to search for other more efficient bacteriocins to improve both safety and shelf-life of meat on the market without altering their organolpetic quality such as enterococcal bacteriocins (enterocins). In this context, further comprehensive research is well recommended to deeply gain insight into enterocins' properties implicated in their production and purification which should be optimised in the near future at economical costs to make their applicability feasible and real in meat industries. Also, the use of enterocins and enterocin-producing *Enterococcus* spp. as bioprotective agents and shelf-life extenders in meat preservation may solve at least a part of the issue regarding practices in meat preservation with chemicals and could thus be more efficient when applied in combination with other technological approaches to greatly reinforce the competitiveness on spoilage and pathogen biocontrol and beneficially enhance food quality. In fact, these suggested advanced methods allow synergistic activities to ensure overall food safety, which meets the consumer-friendly labelling desire.

CONSENT FOR PUBLICATION

Not applicable.

CONFLICT OF INTEREST

The author(s) confirms that there is no conflict of interest.

ACKNOWLEDGEMENTS

Declared none.

REFERENCES

[1] Sharma S. Food preservatives and their harmful effect. IJSR 2015; 5: 1-2.

[2] da Costa RJ, Voloski FLS, Mondadori RG, *et al.* Preservation of meat products with bacteriocins produced by lactic acid bacteria isolated from meat. J Food Qual 2019.
[http://dx.doi.org/10.1155/2019/4726510]

[3] Ahmad V, Khan MS, Jamal QMS, Alzohairy MA, Al Karaawi MA, Siddiqui MU. Antimicrobial potential of bacteriocins: In therapy, agriculture and food preservation. Int J Antimicrob Agents 2017; 49(1): 1-11.
[http://dx.doi.org/10.1016/j.ijantimicag.2016.08.016] [PMID: 27773497]

[4] Orihuel A, Bonacina J, Vildoza MJ, *et al.* Biocontrol of *Listeria monocytogenes* in a meat model using a combination of a bacteriocinogenic strain with curing additives. Food Res Int 2018; 107: 289-96.
[http://dx.doi.org/10.1016/j.foodres.2018.02.043] [PMID: 29580488]

[5] Ghrairi T, Chaftar N, Hani K. Bacteriocins: Recent advances and opportunities. In: Bhat R, Alia AK, Paliyath G, Eds. Progress in food preservation. John Wiley & Sons Ltd 2012; pp. 485-11.

[6] Woraprayote W, Malila Y, Sorapukdee S, Swetwiwathana A, Benjakul S, Visessanguan W. Bacteriocins from lactic acid bacteria and their applications in meat and meat products. Meat Sci 2016; 120: 118-32.
[http://dx.doi.org/10.1016/j.meatsci.2016.04.004] [PMID: 27118166]

[7] Kročko M, Čanigová M, Ducková V. Occurrence, isolation and antibiotic resistance of *Enterococcus* species isolated from raw pork, beef and poultry. J Food Nutr Res 2007; 46: 91-5.

[8] Foulquié Moreno MR, Sarantinopoulos P, Tsakalidou E, De Vuyst L. The role and application of enterococci in food and health. Int J Food Microbiol 2006; 106(1): 1-24.
[http://dx.doi.org/10.1016/j.ijfoodmicro.2005.06.026] [PMID: 16216368]

[9] Alvarez-Cisneros Y, Sáinz Espuñes T, Wacher C, *et al.* Enterocins: Bacteriocins with applications in the food industry. In: Méndez-Vilas A, Ed. Science against microbial pathogens: communicating current research and technological advances. Badajoz: Formatex Research Center 2011; pp. 1112-23.

[10] Hugas M, Garriga M, Aymerich MT. Functionality of enterococci in meat products. Int J Food Microbiol 2003; 88(2-3): 223-33.
[http://dx.doi.org/10.1016/S0168-1605(03)00184-3] [PMID: 14596994]

[11] Kadri Z, Spitaels F, Cnockaert M, *et al. Enterococcus bulliens* sp. nov., a novel lactic acid bacterium isolated from camel milk. Antonie van Leeuwenhoek 2015; 108(5): 1257-65.
[http://dx.doi.org/10.1007/s10482-015-0579-z] [PMID: 26346480]

[12] Ben Braïek O, Smaoui S. Enterococci: Between emerging pathogens and potential probiotics. BioMed Res Int 2019; 2019: 5938210.
[http://dx.doi.org/10.1155/2019/5938210] [PMID: 31240218]

[13] García-Solache M, Rice LB. The *Enterococcus*: A model of adaptability to its environment. Clin Microbiol Rev 2019; 32(2): e00058-18.
[http://dx.doi.org/10.1128/CMR.00058-18] [PMID: 30700430]

[14] Gaaloul N, Ben Braiek O, Berjeaud JM, *et al.* Evaluation of antimicrobial activity and safety aspect of *Enterococcus italicus* GGN10 strain isolated from Tunisian bovine raw milk. J Food Saf 2014; 34: 300-11.
[http://dx.doi.org/10.1111/jfs.12126]

[15] Javed MA, Masud T, Riaz QA, *et al.* Enterocins of *Enterococcus faecium*, emerging natural food preservatives. Ann Microbiol 2011; 61: 699-08.
[http://dx.doi.org/10.1007/s13213-011-0223-8]

[16] Liu G, Griffiths MW, Wu P, *et al. Enterococcus faecium* LM-2, a multi-bacteriocinogenic strain naturally occurring in 'Byaslag', a traditional cheese of Inner Mongolia in China. Food Control 2011;

22: 283-9.
[http://dx.doi.org/10.1016/j.foodcont.2010.07.023]

[17]   Ben Braïek O, Ghomrassi H, Cremonesi P, *et al.* Isolation and characterisation of an enterocin P-producing Enterococcus lactis strain from a fresh shrimp (*Penaeus vannamei*). Antonie van Leeuwenhoek 2017; 110(6): 771-86.
[http://dx.doi.org/10.1007/s10482-017-0847-1] [PMID: 28265787]

[18]   Chakchouk-Mtibaa A, Elleuch L, Smaoui S, *et al.* An antilisterial bacteriocin BacFL31 produced by *Enterococcus faecium* FL31 with a novel structure containing hydroxyproline residues. Anaerobe 2014; 27: 1-6.
[http://dx.doi.org/10.1016/j.anaerobe.2014.02.002] [PMID: 24583094]

[19]   Ben Braïek O, Smaoui S, Fleury Y, Morandi S, Hani K, Ghrairi T. Bio-guided purification and mass spectrometry characterisation exploring the lysozyme-like protein from *Enterococcus lactis* Q1, an unusual marine bacterial strain. Appl Biochem Biotechnol 2019; 188(1): 43-53.
[http://dx.doi.org/10.1007/s12010-018-2886-0] [PMID: 30311172]

[20]   Ben Braïek O, Merghni A, Smaoui S, Mastouri M. *Enterococcus lactis* Q1 and 4CP3 strains from raw shrimps: Potential of antioxidant capacity and anti-biofilm activity against methicillin-resistant *Staphylococcus aureus* strains. Lebensm Wiss Technol 2019; 102: 15-21.
[http://dx.doi.org/10.1016/j.lwt.2018.11.095]

[21]   Svetoch EA, Eruslanov BV, Levchuk VP, *et al.* [Antimicrobial activity of bacteriocin S760 produced by *Enterococcus faecium* strain LWP760]. Antibiot Khimioter 2011; 56(1-2): 3-9.
[PMID: 21780664]

[22]   Ben Braïek O, Cremonesi P, Morandi S, Smaoui S, Hani K, Ghrairi T. Safety characterisation and inhibition of fungi and bacteria by a novel multiple enterocin-producing *Enterococcus lactis* 4CP3 strain. Microb Pathog 2018; 118: 32-8.
[http://dx.doi.org/10.1016/j.micpath.2018.03.005] [PMID: 29524547]

[23]   Jennes W, Dicks LMT, Verwoerd DJ. Enterocin 012, a bacteriocin produced by *Enterococcus gallinarum* isolated from the intestinal tract of ostrich. J Appl Microbiol 2000; 88(2): 349-57.
[http://dx.doi.org/10.1046/j.1365-2672.2000.00979.x] [PMID: 10736005]

[24]   Wachsman MB, Farías ME, Takeda E, *et al.* Antiviral activity of enterocin CRL35 against herpesviruses. Int J Antimicrob Agents 1999; 12(4): 293-9.
[http://dx.doi.org/10.1016/S0924-8579(99)00078-3] [PMID: 10493605]

[25]   Todorov SD, Wachsman MB, Knoetze H, Meincken M, Dicks LM. An antibacterial and antiviral peptide produced by *Enterococcus mundtii* ST4V isolated from soya beans. Int J Antimicrob Agents 2005; 25(6): 508-13.
[http://dx.doi.org/10.1016/j.ijantimicag.2005.02.005] [PMID: 15869868]

[26]   Todorov SD, Wachsman M, Tomé E, *et al.* Characterisation of an antiviral pediocin-like bacteriocin produced by *Enterococcus faecium.* Food Microbiol 2010; 27(7): 869-79.
[http://dx.doi.org/10.1016/j.fm.2010.05.001] [PMID: 20688228]

[27]   Al Kassaa I, Hober D, Hamze M, Chihib NE, Drider D. Antiviral potential of lactic acid bacteria and their bacteriocins. Probiotics Antimicrob Proteins 2014; 6(3-4): 177-85.
[http://dx.doi.org/10.1007/s12602-014-9162-6] [PMID: 24880436]

[28]   Gilmore MS, Segarra RA, Booth MC, Bogie CP, Hall LR, Clewell DB. Genetic structure of the *Enterococcus faecalis* plasmid pAD1-encoded cytolytic toxin system and its relationship to lantibiotic determinants. J Bacteriol 1994; 176(23): 7335-44.
[http://dx.doi.org/10.1128/JB.176.23.7335-7344.1994] [PMID: 7961506]

[29]   Sawa N, Wilaipun P, Kinoshita S, *et al.* Isolation and characterization of enterocin W, a novel two-peptide lantibiotic produced by *Enterococcus faecalis* NKR-4-1. Appl Environ Microbiol 2012; 78(3): 900-3.
[http://dx.doi.org/10.1128/AEM.06497-11] [PMID: 22138996]

[30]   Ghomrassi H, ben Braiek O, Choiset Y, *et al.* Evaluation of marine bacteriocinogenic enterococci strains with inhibitory activity against fish-pathogenic Gram-negative bacteria. Dis Aquat Organ 2016; 118(1): 31-43.
[http://dx.doi.org/10.3354/dao02953] [PMID: 26865233]

[31]   Fisher K, Phillips C. The ecology, epidemiology and virulence of *Enterococcus.* Microbiology 2009; 155(Pt 6): 1749-57.
[http://dx.doi.org/10.1099/mic.0.026385-0] [PMID: 19383684]

[32]   Aguilar-Galvez A, Dubois-Dauphin R, Campos D, Thonart P. Genetic determination and localization of multiple bacteriocins produced by *Enterococcus faecium* CWBI-B1430 and *Enterococcus mundtii* CWBI-B1431. Food Sci Biotechnol 2011; 20: 289-96.
[http://dx.doi.org/10.1007/s10068-011-0041-6]

[33]   Cintas LM, Casaus P, Håvarstein LS, Hernández PE, Nes IF. Biochemical and genetic characterization of enterocin P, a novel sec-dependent bacteriocin from *Enterococcus faecium* P13 with a broad antimicrobial spectrum. Appl Environ Microbiol 1997; 63(11): 4321-30.
[http://dx.doi.org/10.1128/AEM.63.11.4321-4330.1997] [PMID: 9361419]

[34]   Achemchem F, Martínez-Bueno M, Abrini J, Valdivia E, Maqueda M. *Enterococcus faecium* F58, a bacteriocinogenic strain naturally occurring in Jben, a soft, farmhouse goat's cheese made in Morocco. J Appl Microbiol 2005; 99(1): 141-50.
[http://dx.doi.org/10.1111/j.1365-2672.2005.02586.x] [PMID: 15960674]

[35]   Casaus P, Nilsen T, Cintas LM, Nes IF, Hernández PE, Holo H. Enterocin B, a new bacteriocin from *Enterococcus faecium* T136 which can act synergistically with enterocin A. Microbiology 1997; 143(Pt 7): 2287-94.
[http://dx.doi.org/10.1099/00221287-143-7-2287] [PMID: 9245817]

[36]   Ghrairi T, Frère J, Berjeaud JM, Manai M. Purification and characterisation of bacteriocins produced by *Enterococcus faecium* from Tunisian rigouta cheese. Food Control 2008; 19: 162-9.
[http://dx.doi.org/10.1016/j.foodcont.2007.03.003]

[37]   Migaw S, Ghrairi T, Belguesmia Y, *et al.* Diversity of bacteriocinogenic lactic acid bacteria isolated from Mediterranean fish viscera. World J Microbiol Biotechnol 2014; 30(4): 1207-17.
[http://dx.doi.org/10.1007/s11274-013-1535-6] [PMID: 24189971]

[38]   Ben Braïek O, Morandi S, Cremonesi P, Smaoui S, Hani K, Ghrairi T. Safety, potential biotechnological and probiotic properties of bacteriocinogenic *Enterococcus lactis* strains isolated from raw shrimps. Microb Pathog 2018; 117: 109-17.
[http://dx.doi.org/10.1016/j.micpath.2018.02.021] [PMID: 29438718]

[39]   Rehaiem A, Ben Belgacem Z, Edalatian MR, *et al.* Assessment of potential probiotic properties and multiple bacteriocin encoding-genes of the technological performing strain *Enterococcus faecium* MMRA. Food Control 2014; 37: 343-50.
[http://dx.doi.org/10.1016/j.foodcont.2013.09.044]

[40]   Du L, Liu F, Zhao P, Zhao T, Doyle MP. Characterization of *Enterococcus durans* 152 bacteriocins and their inhibition of *Listeria monocytogenes* in ham. Food Microbiol 2017; 68: 97-103.
[http://dx.doi.org/10.1016/j.fm.2017.07.002] [PMID: 28800831]

[41]   Nilsen T, Nes IF, Holo H. Enterolysin A, a cell wall-degrading bacteriocin from *Enterococcus faecalis* LMG 2333. Appl Environ Microbiol 2003; 69(5): 2975-84.
[http://dx.doi.org/10.1128/AEM.69.5.2975-2984.2003] [PMID: 12732574]

[42]   Moreno MRF, Leisner JJ, Tee LK, *et al.* Microbial analysis of Malaysian tempeh, and characterization of two bacteriocins produced by isolates of *Enterococcus faecium.* J Appl Microbiol 2002; 92(1): 147-57.
[http://dx.doi.org/10.1046/j.1365-2672.2002.01509.x] [PMID: 11849339]

[43]   Klaenhammer TR. Genetics of bacteriocins produced by lactic acid bacteria. FEMS Microbiol Rev

1993; 12(1-3): 39-85.
[http://dx.doi.org/10.1111/j.1574-6976.1993.tb00012.x] [PMID: 8398217]

[44]   Stepper J, Shastri S, Loo TS, *et al.* Cysteine *S*-glycosylation, a new post-translational modification found in glycopeptide bacteriocins. FEBS Lett 2011; 585(4): 645-50.
[http://dx.doi.org/10.1016/j.febslet.2011.01.023] [PMID: 21251913]

[45]   Maky MA, Ishibashi N, Zendo T, *et al.* Enterocin F4-9, a novel O-linked glycosylated bacteriocin. Appl Environ Microbiol 2015; 81(14): 4819-26.
[http://dx.doi.org/10.1128/AEM.00940-15] [PMID: 25956765]

[46]   Nes IF, Diep DB, Håvarstein LS, Brurberg MB, Eijsink V, Holo H. Biosynthesis of bacteriocins in lactic acid bacteria. Antonie van Leeuwenhoek 1996; 70(2-4): 113-28.
[http://dx.doi.org/10.1007/BF00395929] [PMID: 8879403]

[47]   van Belkum MJ, Stiles ME. Nonlantibiotic antibacterial peptides from lactic acid bacteria. Nat Prod Rep 2000; 17(4): 323-35.
[http://dx.doi.org/10.1039/a801347k] [PMID: 11014335]

[48]   Liu W, Pang H, Zhang H, Cai Y. Biodiversity of lactic acid bacteria.Lactic Acid Bacteria. Dordrecht: Springer Netherlands 2014; pp. 103-3.
[http://dx.doi.org/10.1007/978-94-017-8841-0_2]

[49]   Kasra-Kermanshahi R, Mobarak-Qamsari E. Inhibition effect of lactic acid bacteria against food born pathogen, *Listeria monocytogenes.* Appl Food Biotechnol 2015; 2: 11-9.

[50]   Alvarez-Sieiro P, Montalbán-López M, Mu D, Kuipers OP. Bacteriocins of lactic acid bacteria: extending the family. Appl Microbiol Biotechnol 2016; 100(7): 2939-51.
[http://dx.doi.org/10.1007/s00253-016-7343-9] [PMID: 26860942]

[51]   Kumariya R, Garsa AK, Rajput YS, Sood SK, Akhtar N, Patel S. Bacteriocins: Classification, synthesis, mechanism of action and resistance development in food spoilage causing bacteria. Microb Pathog 2019; 128: 171-7.
[http://dx.doi.org/10.1016/j.micpath.2019.01.002] [PMID: 30610901]

[52]   Franz CMAP, van Belkum MJ, Holzapfel WH, Abriouel H, Gálvez A. Diversity of enterococcal bacteriocins and their grouping in a new classification scheme. FEMS Microbiol Rev 2007; 31(3): 293-310.
[http://dx.doi.org/10.1111/j.1574-6976.2007.00064.x] [PMID: 17298586]

[53]   Rahmeh R, Akbar A, Kishk M, *et al.* Characterization of semipurified enterocins produced by *Enterococcus faecium* strains isolated from raw camel milk. J Dairy Sci 2018; 101(6): 4944-52.
[http://dx.doi.org/10.3168/jds.2017-13996] [PMID: 29525307]

[54]   Merzoug M, Mosbahi K, Walker D, Karam NE. Screening of the enterocin-encoding genes and their genetic determinism in the bacteriocinogenic *Enterococcus faecium* GHB21 probiotics and antimicrobial proteins. Probiotics Antimicrob Proteins 2019; 11(1): 325-31.
[http://dx.doi.org/10.1007/s12602-018-9448-1] [PMID: 30027472]

[55]   Lauková A, Kandričáková A, Buňková L, Pleva P, Ščerbová J. Sensitivity to enterocins of biogenic amine-producing faecal enterococci from ostriches and pheasants. Probiotics Antimicrob Proteins 2017; 9(4): 483-91.
[http://dx.doi.org/10.1007/s12602-017-9272-z] [PMID: 28342109]

[56]   Lauková A, Szabóová R, Pleva P, Buňková L, Chrastinová Ľ. Decarboxylase-positive *Enterococcus faecium* strains isolated from rabbit meat and their sensitivity to enterocins. Food Sci Nutr 2016; 5(1): 31-7.
[http://dx.doi.org/10.1002/fsn3.361] [PMID: 28070313]

[57]   Hickey RM, Twomey DP, Ross RP, Hill C. Production of enterolysin a by a raw milk enterococcal isolate exhibiting multiple virulence factors. Microbiology 2003; 149(Pt 3): 655-64.
[http://dx.doi.org/10.1099/mic.0.25949-0] [PMID: 12634334]

[58] Ríos Colombo NS, Chalón MC, Navarro SA, Bellomio A. Pediocin-like bacteriocins: New perspectives on mechanism of action and immunity. Curr Genet 2018; 64(2): 345-51.
[http://dx.doi.org/10.1007/s00294-017-0757-9] [PMID: 28983718]

[59] Martin NI, Sprules T, Carpenter MR, *et al.* Structural characterization of lacticin 3147, a two-peptide lantibiotic with synergistic activity. Biochemistry 2004; 43(11): 3049-56.
[http://dx.doi.org/10.1021/bi0362065] [PMID: 15023056]

[60] Ali L, Goraya MU, Arafat Y, Ajmal M, Chen JL, Yu D. Molecular mechanism of quorum-sensing in *Enterococcus faecalis*: Its role in virulence and therapeutic approaches. Int J Mol Sci 2017; 18(5): 960.
[http://dx.doi.org/10.3390/ijms18050960] [PMID: 28467378]

[61] Drider D, Fimland G, Héchard Y, McMullen LM, Prévost H. The continuing story of class IIa bacteriocins. Microbiol Mol Biol Rev 2006; 70(2): 564-82.
[http://dx.doi.org/10.1128/MMBR.00016-05] [PMID: 16760314]

[62] Cui Y, Zhang C, Wang Y, *et al.* Class IIa bacteriocins: Diversity and new developments. Int J Mol Sci 2012; 13(12): 16668-707.
[http://dx.doi.org/10.3390/ijms131216668] [PMID: 23222636]

[63] Héchard Y, Sahl HG. Mode of action of modified and unmodified bacteriocins from Gram-positive bacteria. Biochimie 2002; 84(5-6): 545-57.
[http://dx.doi.org/10.1016/S0300-9084(02)01417-7] [PMID: 12423799]

[64] Gálvez A, Maqueda M, Martínez-Bueno M, Valdivia E. Permeation of bacterial cells, permeation of cytoplasmic and artificial membrane vesicles, and channel formation on lipid bilayers by peptide antibiotic AS-48. J Bacteriol 1991; 173(2): 886-92.
[http://dx.doi.org/10.1128/JB.173.2.886-892.1991] [PMID: 1702784]

[65] González C, Langdon GM, Bruix M, *et al.* Bacteriocin AS-48, a microbial cyclic polypeptide structurally and functionally related to mammalian NK-lysin. Proc Natl Acad Sci USA 2000; 97(21): 11221-6.
[http://dx.doi.org/10.1073/pnas.210301097] [PMID: 11005847]

[66] Grande Burgos MJ, Pulido RP, Del Carmen López Aguayo M, Gálvez A, Lucas R. The cyclic antibacterial peptide enterocin AS-48: isolation, mode of action, and possible food applications. Int J Mol Sci 2014; 15(12): 22706-27.
[http://dx.doi.org/10.3390/ijms151222706] [PMID: 25493478]

[67] Perez RH, Zendo T, Sonomoto K. Circular and leaderless bacteriocins: Biosynthesis, mode of action, applications, and prospects. Front Microbiol 2018; 9: 2085.
[http://dx.doi.org/10.3389/fmicb.2018.02085] [PMID: 30233551]

[68] Kawai Y, Ishii Y, Arakawa K, *et al.* Structural and functional differences in two cyclic bacteriocins with the same sequences produced by lactobacilli. Appl Environ Microbiol 2004; 70(5): 2906-11.
[http://dx.doi.org/10.1128/AEM.70.5.2906-2911.2004] [PMID: 15128550]

[69] Gong X, Martin-Visscher LA, Nahirney D, Vederas JC, Duszyk M. The circular bacteriocin, carnocyclin A, forms anion-selective channels in lipid bilayers. Biochim Biophys Acta 2009; 1788(9): 1797-803.
[http://dx.doi.org/10.1016/j.bbamem.2009.05.008] [PMID: 19463781]

[70] Davidson PM, Bozkurt Cekmer H, Monu EA, Techathvanan C. The use of natural antimicrobials in food: An overview. In: Taylor TM, Ed. Handbook of Natural Antimicrobials for Food Safety and Quality. Cambridge: Woodhead Publishing 2015; pp. 1-27.
[http://dx.doi.org/10.1016/B978-1-78242-034-7.00001-3]

[71] Davidson PM, Branen AL. Food antimicrobials- an introduction. In: Davidson PM, Sofos JN, Branen AL, Eds. Antimicrobials in food. Boca Raton Forida, USA: CRC Press 2005; pp. 1-10.
[http://dx.doi.org/10.1201/9781420028737.ch1]

[72] Yang SC, Lin CH, Sung CT, Fang JY. Corrigendum: Antibacterial activities of bacteriocins:

application in foods and pharmaceuticals. Front Microbiol 2014; 5: 683-93.
[http://dx.doi.org/10.3389/fmicb.2014.00683] [PMID: 25544112]

[73] Ben Braïek O, Smaoui S, Ennouri K, *et al.* RAPD-PCR characterisation of two *Enterococcus lactis* strains and their potential on *Listeria monocytogenes* growth behaviour in stored chicken breast meats: generalised linear mixed□effects approaches. Lebensm Wiss Technol 2019; 99: 244-53.
[http://dx.doi.org/10.1016/j.lwt.2018.09.053]

[74] Ben Braïek O, Smaoui S, Ennouri K, Hani K, Ghrairi T. Genetic analysis with Random Amplified Polymorphic DNA of the multiple enterocin-producing *Enterococcus lactis* 4CP3 strain and its efficient role on the growth of *Listeria monocytogenes* in raw beef meat. BioMed Res Int 2018; 2018: 5827986.
[http://dx.doi.org/10.1155/2018/5827986] [PMID: 29984239]

[75] Mtibaa AC, Smaoui S, Ben Hlima H, Sellem I, Ennouri K, Mellouli L. Enterocin BacFL31 from a safety *Enterococcus faecium* FL31: Natural preservative agent used alone and in combination with aqueous peel onion (*Allium cepa*) extract in ground beef meat storage. BioMed Res Int 2019; 2019: 4094890.
[http://dx.doi.org/10.1155/2019/4094890] [PMID: 31119168]

[76] Cleveland J, Montville TJ, Nes IF, Chikindas ML. Bacteriocins: Safe, natural antimicrobials for food preservation. Int J Food Microbiol 2001; 71(1): 1-20.
[http://dx.doi.org/10.1016/S0168-1605(01)00560-8] [PMID: 11764886]

[77] Pellissery AJ, Vinayamohan PG, Amalaradjou MAR, Venkitanarayanan K. Spoilage bacteria and meat quality.Meat Quality Analysis: Advanced Evaluation Methods, Techniques, and Technologies. London: Academic press 2020; pp. 307-34.
[http://dx.doi.org/10.1016/B978-0-12-819233-7.00017-3]

[78] Lawrie RA, Ledward DA. Lawrie's meat science. Cambridge: Woodhead Publishing Limited 2006.
[http://dx.doi.org/10.1533/9781845691615]

[79] Jayasena DD, Jo C. Essential oils as potential antimicrobial agents in meats and meat products: A review. Trends Food Sci Technol 2013; 34: 96-108.
[http://dx.doi.org/10.1016/j.tifs.2013.09.002]

[80] Jalal H, Para AP, Ganguly S, *et al.* Chemical residues in meat and meat products: A review. WJPLS 2015; 1: 106-22.

[81] Papuc C, Goran G, Predescu C, Nicorescu V. Mechanisms of oxidative processes in meat and toxicity induced by postprandial degradation products: A Review. Compr Rev Food Sci Food Saf 2017; 16: 96-123.
[http://dx.doi.org/10.1111/1541-4337.12241]

[82] Espinosa-Pesqueira D, Roig-Sagués AX, Hernández-Herrero MM. Screening method to evaluate amino acid-decarboxylase activity of bacteria present in spanish artisanal ripened cheeses. Foods 2018; 7(11): 182.
[http://dx.doi.org/10.3390/foods7110182] [PMID: 30404189]

[83] Callewaert R, Hugas M, De Vuyst L. Competitiveness and bacteriocin production of enterococci in the production of Spanish-style dry fermented sausages. Int J Food Microbiol 2000; 57: 33-42.
[http://dx.doi.org/10.1016/S0168-1605(00)00228-2]

[84] Aymerich MT, Garriga M, Costa S, *et al.* Prevention of ropiness in cooked pork by bacteriocinogenic cultures. Int Dairy J 2002; 12: 239-46.
[http://dx.doi.org/10.1016/S0958-6946(01)00143-1]

[85] Sabia C, de Niederhäusern S, Messi P, Manicardi G, Bondi M. Bacteriocin-producing *Enterococcus casseliflavus* IM 416K1, a natural antagonist for control of *Listeria monocytogenes* in Italian sausages ("cacciatore"). Int J Food Microbiol 2003; 87(1-2): 173-9.
[http://dx.doi.org/10.1016/S0168-1605(03)00043-6] [PMID: 12927720]

[86]   Sparo MD, Confalonieri A, Urbizu L, Ceci M, Bruni SF. Bio-preservation of ground beef meat by *Enterococcus faecalis* CECT7121. Braz J Microbiol 2013; 44(1): 43-9.
[http://dx.doi.org/10.1590/S1517-83822013005000003] [PMID: 24159282]

[87]   Lauková A, Czikková S, Laczková S, Turek P. Use of enterocin CCM 4231 to control Listeria monocytogenes in experimentally contaminated dry fermented Hornád salami. Int J Food Microbiol 1999; 52(1-2): 115-9.
[http://dx.doi.org/10.1016/S0168-1605(99)00125-7] [PMID: 10573399]

[88]   Aymerich T, Artigas MG, Garriga M, Monfort JM, Hugas M. Effect of sausage ingredients and additives on the production of enterocin A and B by Enterococcus faecium CTC492. Optimization of *in vitro* production and anti-listerial effect in dry fermented sausages. J Appl Microbiol 2000; 88(4): 686-94.
[http://dx.doi.org/10.1046/j.1365-2672.2000.01012.x] [PMID: 10792528]

[89]   Aymerich T, Garriga M, Ylla J, Vallier J, Monfort JM, Hugas M. Application of enterocins as biopreservatives against *Listeria innocua* in meat products. J Food Prot 2000; 63(6): 721-6.
[http://dx.doi.org/10.4315/0362-028X-63.6.721] [PMID: 10852564]

[90]   Vignolo G, Palacios J, Farías ME, *et al.* Combined effect of bacteriocins on the survival of various *Listeria* species in broth and meat system. Curr Microbiol 2000; 41(6): 410-6.
[http://dx.doi.org/10.1007/s002840010159] [PMID: 11080390]

[91]   Ananou S, Garriga M, Jofré A, *et al.* Combined effect of enterocin AS-48 and high hydrostatic pressure to control food-borne pathogens inoculated in low acid fermented sausages. Meat Sci 2010; 84(4): 594-600. a
[http://dx.doi.org/10.1016/j.meatsci.2009.10.017] [PMID: 20374829]

[92]   Ananou S, Baños A, Maqueda M, *et al.* Effect of combined physico-chemical treatments based on enterocin AS-48 on the control of *Listeria monocytogenes* and *Staphylococcus aureus* in a model cooked ham. Food Control 2010; 21: 478-86. b
[http://dx.doi.org/10.1016/j.foodcont.2009.07.010]

[93]   Turgis M, Stotz V, Dupont C, *et al.* Elimination of *Listeria monocytogenes* in sausage meat by combination treatment: Radiation and radiation-resistant bacteriocins. Radiat Phys Chem 2012; 81: 1185-8.
[http://dx.doi.org/10.1016/j.radphyschem.2012.02.021]

[94]   Islam MS, Husna AA, Islam MA, Khatun MM. Prevalence of *Listeria monocytogenes* in beef, chevon and chicken in Bangladesh. Am J Food Sci Health 2016; 2: 39-44.

[95]   Lennox JA, Etta PO, John GE, Henshaw EE. Prevalence of *Listeria monocytogenes* in fresh and raw fish, chicken and beef. J Adv Microbiol 2017; 3: 1-7.
[http://dx.doi.org/10.9734/JAMB/2017/33132]

[96]   Barroso I, Maia V, Cabrita P, Martínez-Suárez JV, Brito L. The benzalkonium chloride resistant or sensitive phenotype of *Listeria monocytogenes* planktonic cells did not dictate the susceptibility of its biofilm counterparts. Food Res Int 2019; 123: 373-82.
[http://dx.doi.org/10.1016/j.foodres.2019.05.008] [PMID: 31284989]

[97]   Ananou S, Maqueda M, Martínez-Bueno M, Gálvez A, Valdivia E. Control of *Staphylococcus aureus* in sausages by enterocin AS-48. Meat Sci 2005; 71(3): 549-56.
[http://dx.doi.org/10.1016/j.meatsci.2005.04.039] [PMID: 22060932]

[98]   Teixeira LM, Merquior VLC. Enterococcus.Molecular Typing in Bacterial Infections. Totowa, USA: Humana Press Inc 2013; pp. 17-26.
[http://dx.doi.org/10.1007/978-1-62703-185-1_2]

[99]   O'Driscoll T, Crank CW. Vancomycin-resistant enterococcal infections: Epidemiology, clinical manifestations, and optimal management. Infect Drug Resist 2015; 8: 217-30.
[PMID: 26244026]

[100]   Economou V, Sakkas H, Delis G, Gousia P. Antibiotic resistance in *Enterococcus* spp. friend or foe? In: Singh OV, Ed. Foodborne Pathogens and Antibiotic Resistance. JohnWiley & Sons Inc 2017.
[http://dx.doi.org/10.1002/9781119139188.ch16]

[101]   Ladero V, Calles-Enríquez M, Fernández M, Alvarez MA. Toxicological effects of dietary biogenic amines. Curr Nutr Food Sci 2010; 6: 145-56.
[http://dx.doi.org/10.2174/157340110791233256]

[102]   Benkerroum N. Biogenic amines in dairy products: Origin, incidence, and control means. Compr Rev Food Sci Food Saf 2016; 15: 801-26.
[http://dx.doi.org/10.1111/1541-4337.12212]

[103]   Costa MP, Rodrigues BL, Frasao BS, Conte-junior CA. Chemical Risk for Human Consumption. Amsterdam, The Netherlands: Elsevier Inc 2018.

[104]   Aguilar-Galvez A, Dubois-Dauphin R, Destain J, *et al.* Les entérocoques: avantages et inconvénients en biotechnologie (synthèse bibliographique). Biotechnol Agron Soc Environ 2012; 61: 67-76.

[105]   Leuschner R, Kurihara R, Hammes W. Formation of biogenic amines by proteolytic enterococci during cheese ripening. J Sci Food Agric 1999; 79: 1141-4.
[http://dx.doi.org/10.1002/(SICI)1097-0010(199906)79:8<1141::AID-JSFA339>3.0.CO;2-0]

[106]   Burdychova R, Komprda T. Biogenic amine-forming microbial communities in cheese. FEMS Microbiol Lett 2007; 276(2): 149-55.
[http://dx.doi.org/10.1111/j.1574-6968.2007.00922.x] [PMID: 17956420]

[107]   Kučerová K, Svobodová H, Tůma S, *et al.* Production of biogenic amines by enterococci. Czech J Food Sci 2009; 27: 2-50.

[108]   Gardini F, Bover-Cid S, Tofalo R, *et al.* Modeling the aminogenic potential of *Enterococcus faecalis* EF37 in dry fermented sausages through chemical and molecular approaches. Appl Environ Microbiol 2008; 74(9): 2740-50.
[http://dx.doi.org/10.1128/AEM.02267-07] [PMID: 18296537]

[109]   Garriga M, Aymerich MT, Costa S, *et al.* Bactericidal synergism through bacteriocins and high pressure in a meat model system during storage. Food Microbiol 2002; 19: 509-18.
[http://dx.doi.org/10.1006/fmic.2002.0498]

[110]   Jofré A, Aymerich T, Monfort JM, Garriga M. Application of enterocins A and B, sakacin K and nisin to extend the safe shelf-life of pressurized ready-to-eat meat products. Eur Food Res Technol 2008; 228: 159-62.
[http://dx.doi.org/10.1007/s00217-008-0913-z]

[111]   Marcos B, Jofré A, Aymerich T, *et al.* Combined effect of natural antimicrobials and high pressure processing to prevent *Listeria monocytogenes* growth after a cold chain break during storage of cooked ham. Food Control 2008; 19: 76-81.
[http://dx.doi.org/10.1016/j.foodcont.2007.02.005]

[112]   Jofré A, Aymerich T, Garriga M. Improvement of the food safety of low acid fermented sausages by enterocins A and B and high pressure. Food Control 2009; 20: 179-84.
[http://dx.doi.org/10.1016/j.foodcont.2008.04.001]

[113]   Weiss J, Zhong Q, Harte F, Davidson PM. Micro and nanoparticles for controlling microorganisms in foods. In: Pabst G, Kucerka N, Nieh MP, Katsaras J, Eds. Liposomes, lipid bilayers and model membranes: From basic research to technology. Boca Raton, Florida, USA: CRC Press 2014; pp. 415-50.
[http://dx.doi.org/10.1201/b16617-24]

[114]   Marcos B, Aymerich T, Monfort JM, Garriga M. Use of antimicrobial biodegradable packaging to control *Listeria monocytogenes* during storage of cooked ham. Int J Food Microbiol 2007; 120(1-2): 152-8.
[http://dx.doi.org/10.1016/j.ijfoodmicro.2007.06.003] [PMID: 17629977]

[115] Iseppi R, Pilati F, Marini M, *et al.* Anti-listerial activity of a polymeric film coated with hybrid coatings doped with Enterocin 416K1 for use as bioactive food packaging. Int J Food Microbiol 2008; 123(3): 281-7.
[http://dx.doi.org/10.1016/j.ijfoodmicro.2007.12.015] [PMID: 18262299]

[116] Sidhu PK, Nehra K. Bacteriocin-nanoconjugates as emerging compounds for enhancing antimicrobial activity of bacteriocins. J King Saud Univ – Sci 2019; 31: 758-67.

[117] Lopes NA, Brandelli A. Nanostructures for delivery of natural antimicrobials in food. Crit Rev Food Sci Nutr 2018; 58(13): 2202-12.
[http://dx.doi.org/10.1080/10408398.2017.1308915] [PMID: 28394691]

[118] Fahim HA, Khairalla AS, El-Gendy AO. Nanotechnology: A valuable strategy to improve bacteriocin formulations. Front Microbiol 2016; 7: 1385.
[http://dx.doi.org/10.3389/fmicb.2016.01385] [PMID: 27695440]

[119] Sharma TK, Sapra M, Chopra A. Interaction of bacteriocin-capped silver nanoparticles with food pathogens and their antibacterial effect. Int J Green Nanotechnol 2012; 4: 93-110.
[http://dx.doi.org/10.1080/19430892.2012.678757]

CHAPTER 3

Technological Advancement in the Detection and Identification of Plant Pathogens

Yaakoub Gharbi*, **Emna Bouazizi**, **Manel Cheffi** and **Mohamed Ali Triki**

Laboratoire Les Ressources Génétiques de l'Olivier: Caractérisation, Valorisation et Protection Phytosanitaire, Institut de l'Olivier, Université de Sfax, Tunisie

Abstract: Severe yield losses due to crop infections with pathogens such as bacteria, viruses and fungi are challenging issues in agriculture for decades across the world. To limit the impact of disease damage in economically important crops and to ensure agricultural sustainability, developing new diagnostic methods for the rapid and accurate detection of plant pathogens is essential as it helps to prevent major yield losses and preserve a good quality of products at the postharvest stage. In this context, serological techniques such as ELISA and molecular protocols based on PCR, quantitative PCR, isothermal amplification, microarrays and RNA-Seq-based next-generation sequencing are leading to more accurate detection for the most destructive plant pathogens, which reduced the economic losses due to plant disease infections. Despite their reliability in the design of an efficient management program for several plant diseases, the performance of these techniques is sometimes limited by, the unknown distribution of the studied disease, the existence of asymptomatic infections and the lack of validated sampling protocols. Recently, more sophisticated techniques such as thermography, fluorescence imaging, hyper-spectral techniques and biosensors relying on various parameters such as morphological change, temperature change, transpiration rate change and bio-recognition elements such as enzyme, volatile organic compounds, antibody, and DNA/RNA released by infected plants have been applied, either for on-site diagnostic or for detecting plant diseases over large areas. This review briefly describes the various techniques used for plant disease diagnosis and their evolution to meet the contemporary challenges.

Keywords: Biosensors, Chlorophyll fluorescence, Hyperspectral imaging, Nucleic acid-based methods, Plant disease, Pathogen detection, Serological based assays, Thermography.

INTRODUCTION

Plant diseases reduce the world's agricultural productivity by up to 40% yearly,

* **Corresponding author Gharbi Yaakoub:** Laboratoire Les Ressources Génétiques de l'Olivier: Caractérisation, Valorisation et Protection Phytosanitaire Institut de l'Olivier, Université de Sfax, Tunisie; E-mail: yaakoub.gharbi@yahoo.com

Karim Ennouri (Ed.)

leading to potential economic losses and substantial environmental effects from chemical management practices [1, 2]. Thereby, monitoring and early detection of harmful pathogens are essential to prevent their spread and ease the design of successful management programs.

The accurate diagnosis of plant diseases relies on the good knowledge of the biological basis of the host-pathogen interaction and the visual specific symptoms that may develop during infection [3]. Basically, the diagnosis of crop diseases is based on assessing the visual symptoms and their severity using traditional disease scales [4]. However, these methods are too subjective, due to the similarity of symptoms and changes in the host tissue morphology caused by biotic or abiotic stress [3, 4]. In addition, visual inspection does not allow discrimination of diseased plants from symptomless ones, which underestimates the threat posed by the pathogen and hampers the predicting of possible disease outbreak onset [3, 5]. Therefore, effective management programs against harmful pathogens have to be performed before the onset of disease outbreak, which requires an accurate and specific diagnosis during the early stage of infection [6].

Over the last two decades, detection of plant pathogens has shifted from the visual inspection of symptoms, towards molecular and immunological approaches. Several protocols relying on the Enzyme Linked Immunosorbent Assay (ELISA) and Polymerase Chain Reaction (PCR) were successfully used to resolve several cases of biotic infections [7 - 10]. The ELISA assay, which is the most used serological method, was first optimized for detecting plant viruses [11] and then applied for detecting fungal and bacterial pathogens using monoclonal and polyclonal antisera [12 - 14]. However, due to their low sensitivity and specificity, serological methods are likely unsuitable for large phytosanitary inspections. Recently, molecular DNA-based technologies have come into play with higher specificity and sensitivity, which allow detecting a low pathogen titer in the host tissues [15, 16]. The PCR-based methods targeting genus- or species- specific DNA markers have been widely employed for identifying the most harmful plant pathogens [17, 18]. Several molecular approaches such as Reverse Transcriptase-PCR (RT-PCR), Nucleic acid sequence-based amplification (NASBA), and other PCR variants (conventional PCR, Nested PCR, Multiplex PCR, and Real-Time PCR) were developed for detecting pathogens in plants and environmental samples [3, 9, 10]. In fact, molecular techniques have been successfully used to discriminate morphologically similar species, detect asymptomatic infection, and conduct large-scale epidemiological studies [19 - 22]. Nowadays, more sophisticated PCR based protocols have been designed to be rapid, quantitative with on-site diagnostic features. As an example, the real-time PCR method is designed to yield amplicons from a target region of a pathogen's genome using specific primers and fluorescent probes [23]. This technique quantifies the pathogens that cannot be isolated or cultured easily from plant tissue, or present at low concentration in the diagnosed samples [23, 24]. Portable real-time PCR instruments are now available for on-site diagnostic assays under field conditions other than the conventional diagnostic laboratory conditions [25]. The Loop Mediated Isothermal Amplification (LAMP) method was also developed to be directly applied in the field, which is reliable in large epidemiological studies and improve decision-making in disease control (Fig. **1**) [26].

Recently, non-destructive approaches, including hyperspectral imaging, thermal imaging and biosensors were developed and applied for large-scale monitoring of many plant diseases and their on-site detection (Fig. **1**) [27]. Thermography, fluorescence imaging, hyperspectral sensors and gas chromatography are the most promising technologies [28]. Actually, biosensors may check the optical features of plants within different regions of the electromagnetic spectrum, to detect early anatomical and physiological changes such as, tissue color variation, leaf shape, transpiration rate, and canopy morphology [29]. However, using spectral approaches is a matter of debate in plant pathology, due to the similarity of symptoms caused by environmental factors and those caused by biotic diseases [30]. Further, use of image analysis is rapidly increasing in the detection of the most plant pathogens, threatening agriculture production worldwide [31, 32].

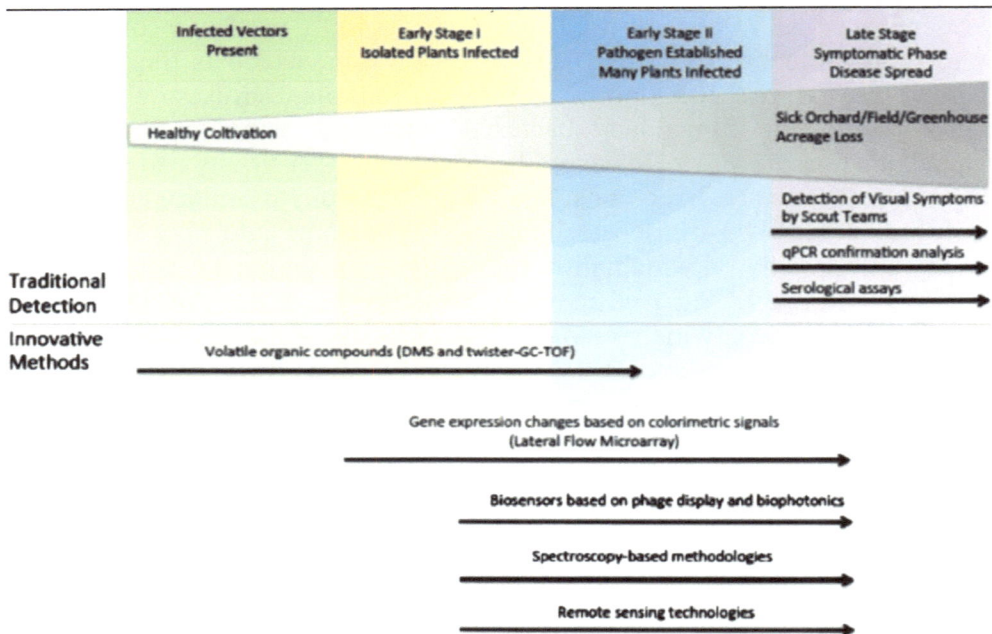

Fig. (1). Timing for useful application of traditional and innovative techniques adopted for plant disease diagnosis [86].

This book chapter provides a comprehensive review of the techniques adopted for plant pathogen detection and discusses the advanced methods applied for early identification of crop diseases.

SEROLOGICAL BASED METHODS

Enzyme Linked Immunosorbent Assay

Serological assays were first developed to detect uncultured microorganisms [11]. The ELISA assay relies on the specific recognition of an immobilized antigen by an antibody linked to an enzyme. The detection is then achieved by measuring the conjugated enzyme activity during incubation with a substrate, which generates a measurable product (Fig. **2**). Actually, several bacterial and fungal pathogens are specifically detected using polyclonal and monoclonal antisera and techniques such as ELISA, western blot, and dot-blot immune-binding assays [3]. So far, ELISA is the most widely used technique because of its high throughput potential and its reliability to conduct large-scale epidemiological studies [33]. However, the detection limit of this technique varies depending on the sample type and preparation, and the pathogen titer level [34]. ELISA rapid detection kits using both poly and monoclonal antibodies are commercially available for several plant pathogens. For instance, many ELISA kits were optimized for detecting bacterial pathogens such as *Pseudomonas syringae* [35], *Ralstonia solanacearum* [36], *Erwinia amylovora* [37], *Xylella fastidiosa* [14], *Dickeya dadantii* [38], and *Pectobacterium carotovorum* [39]. Also, ELISA assays have been also applied to detect fungal pathogens such as *Verticillium dahliae* and *Botrytis cinerea*. Although the great improvement of ELISA assays accuracy, their frequent cross-reactivity inspired the development of more effective monoclonal antisera using hybridoma technology with specific cell lines to single epitopes [40]. To make the diagnosis practically more reliable, some ELISA kits are even applicable in the field or at any part of the production and distribution system. For instance, the lateral flow-ELISA is widely employed for *Phytophthora infestans*, *Ralstonia solanacearum*, tomato mosaic virus, the potyvirus group, and many other pathogens [41]. Similarly, tissue print-ELISA and other lateral flow devices are also used for on-site specific detection [42].

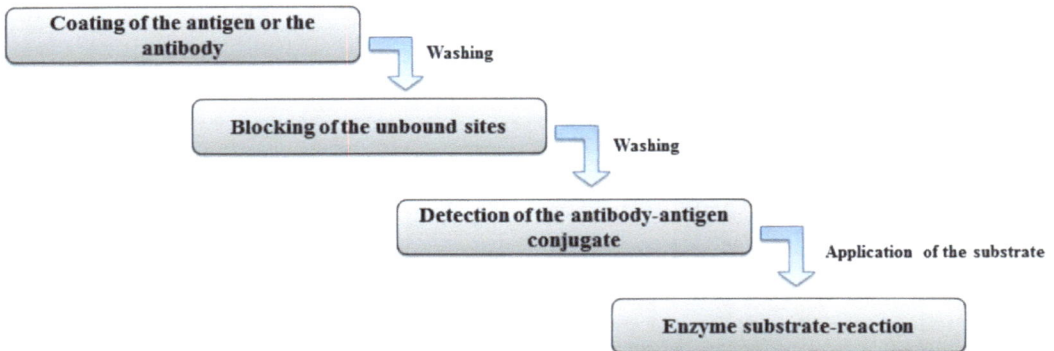

Fig. (2). A general scheme representing the major steps of an ELISA assay.

Immunofluorescence

The immunofluorescence (IF) is a microscopy-based optical technique used for analyzing microbiological samples. In this technique, specific antibodies bound to their target antigens are detected by using second antibodies conjugated with fluorescent dyes such as, fluorescein isothiocyanate or rhodamine isothiocyanate. Subsequently, the fluorescence signalling the presence of the target antigen is visualized by the microscope. Immunofluorescence may be useful for detecting fungal spores, which are often too large to bind to ELISA microplates without prior disruption. The technique can also be used to detect pathogenic bacteria in plant tissues [43]. IF was used to detect the fungus *Phakopsora pachyrhizi* causing rust disease in soybean and *Clavibacter michiganensis subsp. sepedonicus* causing bacterial ring rot in potatoes [44, 45]. However, the major drawback of IF is the equipment availability, the lack of quantification, and technical issues such as, autofluorescence of plant tissue or soil particles.

NUCLEIC ACID-BASED METHODS

Since the development of PCR in the 1980s, several variants of this technique were optimized for detecting serious pathogens in plant pathology (Table 1). PCR based methods could be classified into two major groups. Some protocols are DNA based such as conventional PCR, nested PCR, multiplex PCR, immune-capture PCR, and real-time PCR, whereas others are RNA based such as RT-PCR and NASBA [5].

DNA-based PCR Methods

These methods are suitable for direct detection of pathogens from host tissues without their isolation in culture media (Fig. **3**). DNA-based PCR are more specific and time-saving than the processes of conventional methods, which are unspecific and show many contaminants that may confuse the identification step [3, 17]. In addition, DNA-based PCR are more sensitive and allow detecting low pathogen titer, which is not allowed by conventional methods [17, 46]. The other advantage of the DNA based methods is their utility to compare the effect of different treatments such as the effect of fungicide application on the incidence rate of plant diseases [47].

Table 1. Advantages and limitations of the available molecular-based techniques adopted for detecting plant pathogens.

Molecular methods	Advantages	Disadvantages
Conventional PCR	- Rapid diagnosis, with high specificity when specific primers and right PCR conditions are used	- Susceptible to enzymatic inhibitors present in plant and insect tissues - High risk of contamination
Multiplex PCR	- Simultaneous detection of more than one pathogen - Time saving and cost-effective	- Specificity of the reaction depends on the primers designed - Risk of cross amplification and formation of primer dimers
Nested PCR	- Effective for detection of low pathogens titer - detection of pathogens in symptomless plant tissues	- High risk of contamination when performed at two consecutive amplification steps - difficulty to find compatible primers with no dimers formation
Multiplex Nested PCR	- Effective for detection of low pathogens titer - detection of pathogens in symptomless plant tissues -Time saving and cost effective	- high risk of contamination when performed at two consecutive amplification steps - Critical optimization of PCR conditions (annealing temperature, annealing time, extension time)
IC-PCR	- Effective for detection of pathogens present with low titer in plant tissue	- required special expertise - Higher cost compared to conventional PCR
Real time PCR	- Highly sensitive and quantitative - low risk of contamination	- Higher cost when compared to conventional PCR - need selection of specific primers targeting DNA sequences with no more 200 bp
Reverse Transcriptase (RT)-PCR	- Higher sensitivity and specificity - detect viable cells of the pathogen - low risk of contamination	- requires RNA isolation - Higher cost compared to real time PCR - required special expertise
LAMP	- Rapid, sensitive and highly specific	- difficulty of primer design - risk of contamination
RNA-Seq	- Increased specificity and sensitivity	- Needs expensive equipment - Strong bio-informatics knowledge to process and analyze the data

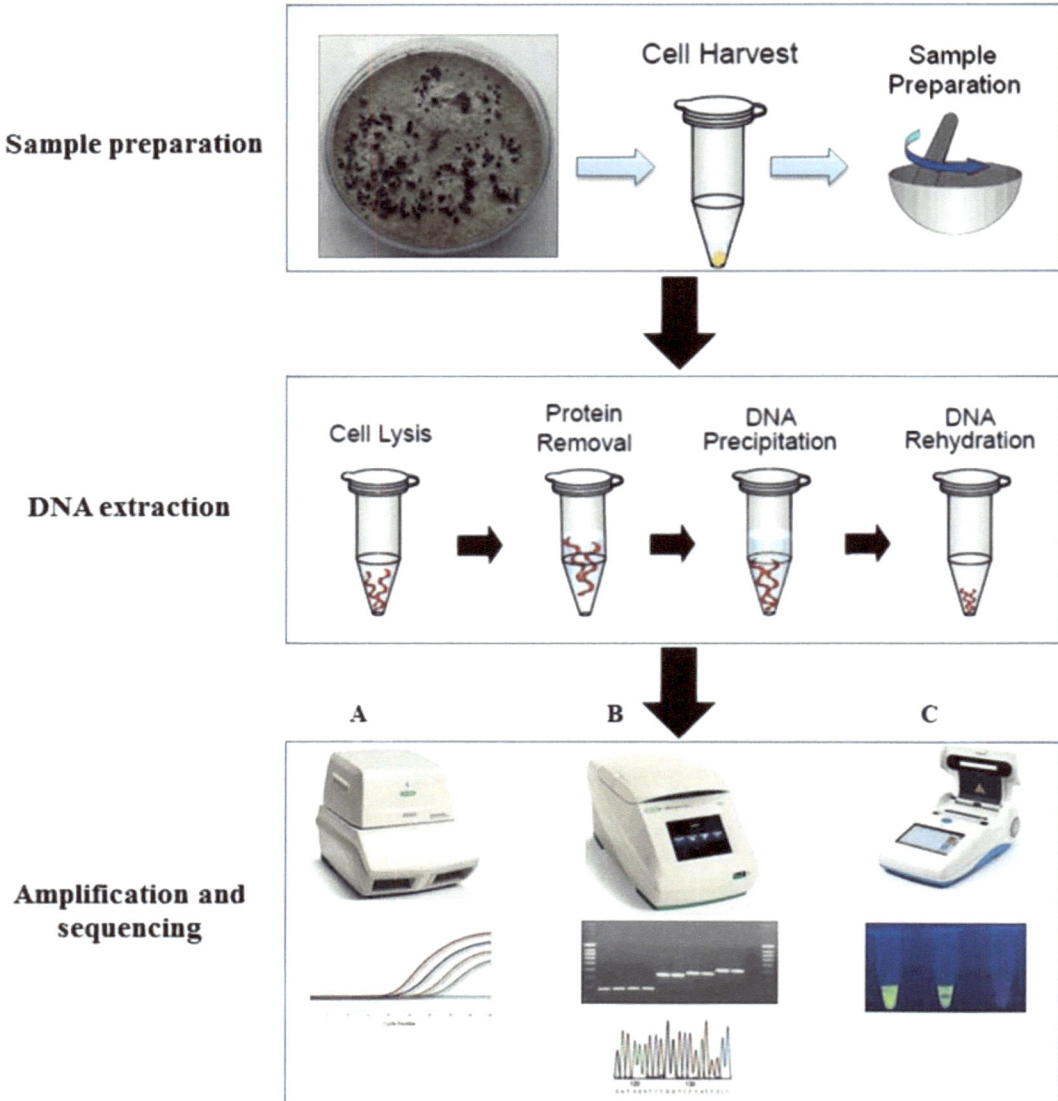

Fig. (3). An overview of the general steps for the molecular identification of plant pathogens. (**A**: real time PCR; **B**: Conventional PCR and DNA sequencing; **C**: LAMP PCR).

Multiplex PCR, in which primers for two or more targets can be conjointly used to detect different pathogens in a single closed tube [48]. This technique is largely employed to assess infection caused simultaneously by two or more pathogens [49]. For instance, multiplex PCR was successfully employed to diagnose viral diseases in citrus [50] and tomato plants [51]. Nested PCR involves two sets of primers in two successive reactions, with amplicons from the first round used as a template for the second one. This technique exhibits greater sensitivity than conventional PCR, which is useful for detecting low pathogen titer [52]. Nested PCR was used to detect *Pilidiella granati* in pomegranate fruit [53], *Colletotrichum acutatum* on Symptomless Strawberry Leaves [54], and *Mycosphaerella* spp. on Eucalyptus leaves [55]. Also, nested PCR was useful for detecting uncultured phytoplasmas in plant host tissue [56]. However, the use of nested PCR favors risk of self-contamination when preparing the second round of amplification. To avoid this problem, nested PCR in a single closed tube has been adopted [57]. A new PCR variant called multiplex nested PCRwas used to detect several pathogens in a single reaction [58] or genetically heterogeneous strains of a single pathovar [59]. This technique was useful for discriminating different VCG groups of *Verticillium dahliae* [60]. It was also used for detecting *Xanthomonas axonopodis* pv. *allii* in onion and other *Allium* species [61].

A new technique called immune-capture PCR, in which immobilized antibodies are used for specific detection of pathogen cells and ease their further detection by PCR. This approach improves the PCR sensitivity and alleviates the impact of enzyme inhibitors present in the tested samples [62]. The immune-capture PCR is a useful alternative in virus detection in plant material [63].

Quantifying the pathogen DNA in host tissues is important for assessing cultivars susceptibility to such disease [64], and for evaluating the effect of applied pest management programs [65]. However, this parameter can't be studied using conventional PCR, since this later is inherently not quantitative. The real time PCR, which is a newer technology exhibits higher sensitivity and rapidity than conventional PCR, due to many factors, including the amplification of small DNA sequences, which allow the finding of specie-specific fragments and therefore the design of specific probes [66]. The principle of detection by Real time PCR is mainly based on using different types of fluorescent dyes, which allows the quantification of small amounts of amplified products, allowing detection of symptomless diseases [67]. Furthermore, the assay is performed in a closed tube without electrophoresis separation and gel staining, which reduces post-PCR contamination and false positive results [68]. In real-time PCR, there are two common approaches for detecting amplified products, which differ by the type of the fluorescent dye. First, non-specific fluorescent dyes that intercalate with any double-stranded DNA such as SYBR [68]. Secondly, specific DNA probes consisting of oligo-nucleotides labeled with a fluorescent reporter, which emit signal only after hybridization with its complementary sequence [67]. Therefore, these alternative detection chemistries make real-time PCR more suitable for multiplexing detection purposes and promote the simultaneous assay of more than one pathogen [69]. Identifying a specie-specific target for primer design is a critical step towards successful pathogen detection. In this context, many DNA target such as the genes coding for β-tubulin and mitochondrial small subunit rRNA have been used as targets in real time PCR for fungal detection (Table **2**). The detection limits of real time PCR reported in previous studies is ranging between 10 pg/mL and 10 fg/mL of initial DNA amount [67, 68]. In addition, real time PCR usually amplify short DNA sequences, which increase PCR specificity and sensitivity compared to conventional PCR [68]. However, using short target is not recommended thanks to their confusion with primer dimers when resolved in agarose gels. Several research studies proved the reliability of real time PCR not only for pathogen quantification, but also for determining the resistance/tolerance traits of plant species to some diseases. In fact, this technique was successfully used to discriminate between susceptible and tolerant olive cultivars to *Verticillium dahliae* [64, 70], and evaluation of resistant and susceptible Hop cultivars in response to infection with *Verticillium albo-atrum* [71]. In fact, quantifying soil-borne pathogens is a critical parameter for assessing the phytosanitary status of soils before implementing new plantations [65, 72]. Real time PCR was successfully used to assess the phytosanitary status of soils infected by the soilborne pathogens *Verticillium dahliae* [65] and *Fusarium oxysporum* [72]. Bacterial pathogens have been also quantified and detected in host tissues (Table **1**). The multiplexing option is also available in Real-Time PCR allowing the simultaneous detection of more than one pathogen using probes with different fluorescent reporter dyes. In this context, a multiplex real-time PCR targeting a conserved DNA region with a single-nucleotide polymorphism mutation was adopted to detect and identify the bacterial spot disease caused by different *Xanthomonas* species [73]. Real time PCR is also useful to survey the efficacy of applied management programs aiming to reduce fungal inoculum in heavily infected soils.

Table 2. Real Time-Based PCR assays for detecting fungal and bacterial pathogens infecting important commercial crops in the world.

Pathogen	Pathogen specie	Host	Chemistry	Reference
Fungal pathogens	*Fusarium oxysporum*	Chickpea	SYBR Green	[66]
	Macrophomina phaseolina	Chickpea, soybean	TaqMan minor groove binder (MGB)	[16]
	Rhizoctonia oryzae	Cereals	SYBR Green	[22]
	Verticillium dahliae	Potato, Olive	SYBR Green	[15, 70]
	Pythium irregulare	Wheat, Barley	SYBR Green	[75]
	Phytophthora erythroseptica	Potato	TaqMan	[74]

Pathogen	Pathogen specie	Host	Chemistry	Reference
Bacterial pathogens	*Erwinia uzenensis*	Pear	TaqMan	[76]
	Xylella fastidiosa	Grapevine, Oak	TaqMan	[77]
	Erwinia amylovora	Pear, Apple, Quince	SYBR Green	[78]

LAMP is a sensitive and specific tool for detecting pathogens by combination of specific primers coupled with a fluorescent dye, which enhance the detection limit of this technique and its ability to screen plants with asymptomatic infections [79]. Another advantage of LAMP is its usefulness for field diagnosis, especially when large-scale analysis is essential in contaminated area under quarantine regulations [80]. However, LAMP protocols need to be validated before their admission for official analysis, by testing them under different conditions such as, the pathogen specie and the type of analyzed tissue [81]. LAMP was recently adopted for monitoring *Xylella fastidiosa* in olive and grapevine [81]. LAMP could be employed to conduct large epidemiological studies and disease surveys in contaminated areas.

RNA-BASED METHODS

Though their high sensitivity, DNA-based PCR methods are enable to discriminate between living and dead cells. In fact, assessing cells viability is important for evaluating the treatments intended to reduce pathogen inoculums and/or limit disease symptoms progression. The RT-PCR target mRNA molecules as their presence in the sample could be considered an accurate sign of cell viability [82]. In RT-PCR assay, the mRNA sequences are reverse transcribed with the reverse transcriptase enzyme to produce cDNA molecules after the enzymatic reaction (Fig. **4**). The cDNA is then amplified by conventional or real time PCR. Although, real time PCR is frequently applied for analyzing the gene expression of the host or the pathogen, it is also adopted to detect viable inoculums of fungal pathogens such as *Mycosphaerella graminicola, Pyrenophora tritici-repentis and Parastagonospora nodorum* in wheat [83, 84]. RT-PCR has been also used to quantify *Fusarium graminearum* causal agent of ear blight disease in wheat [85]. Finally, a multiplex nested RT-PCR in a single tube for sensitive and simultaneous detection of three RNA viruses infecting *Narcissus* spp [86].

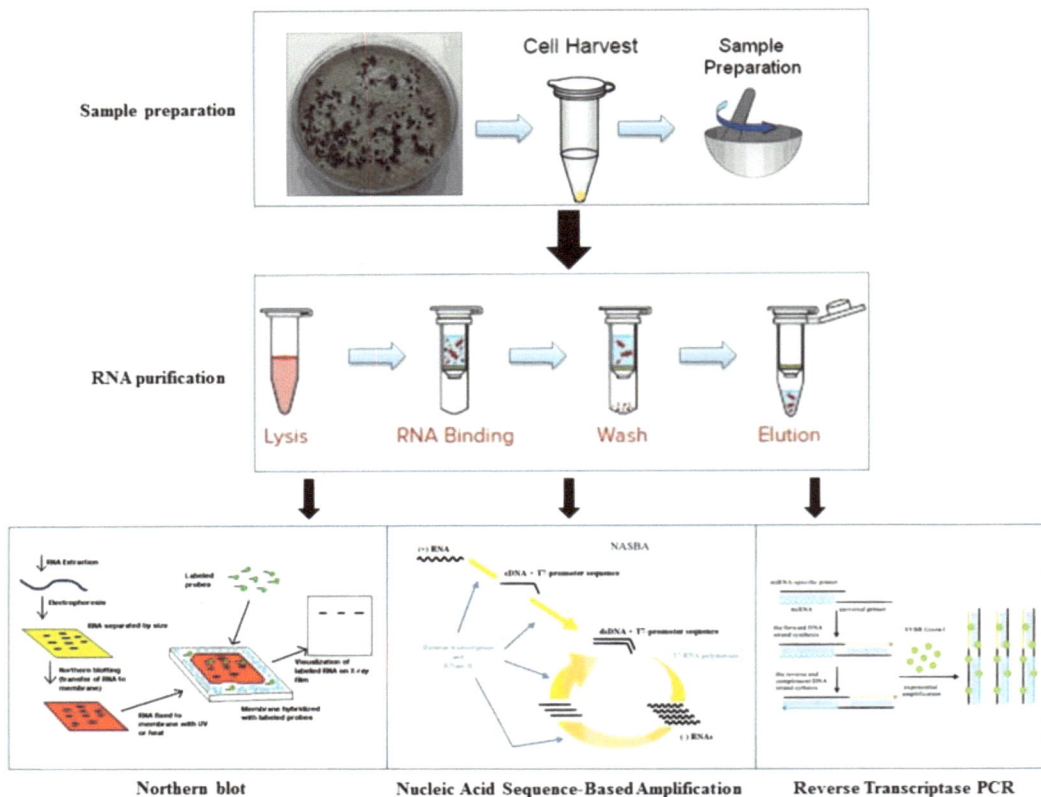

Fig. (4). An overview of the general steps for the molecular identification of plant pathogens using RNA based methods.

Northern blot, also called RNA blot, adopted the same concept as the Southern blot except the RNA material is used instead of the DNA. RNA purification from plant tissue is the first step that should be performed with maximum care in order to achieve high quality RNA with high molecular weight. Subsequently, the purified RNA is separated in agarose gel electrophoresis, which is then blotted on a special filter paper. The filter is then exposed to radioactive labeled probe for eventual hybridization with its complementary RNA sequence. Autoradiography is then used to visualize the film and the presence of a band indicates a positive reaction. The Northern blot is useful to study the gene expression although RT-PCR is more reliable in this context [87]. However, this technique has some limits such as the risk of mRNA denaturation during the electrophoresis process. Also, the sensitivity of northern blotting technique is relatively low compared to the RT-PCR. Further, this technique is not suitable to conduct large scale epidemiological studies. Also, the protocol is complex and time consuming [87].

This technique was applied to detect *Magnaporthe grisea* infecting rice by using the real-time PCR and Northern Blot [88].

NASBA is an isothermal reaction that is performed at 41°C. This technique has the particularity to amplify RNA directly with high sensitivity unlike northern blot. NASBA has the advantage to amplify more than 10^9 copies of the target RNA sequence in a short time without using thermal cycler devices [84]. Previous studies proved that NASBA exhibit comparable sensitivity as RT-PCR [89]. NASBA has the advantage of producing single-stranded RNA amplicons that can be used directly in another round of amplification [89]. However, the big limit of NASBA is the quality of purified RNA and its integrity [90]. Also, the specificity of NASBA assay is influenced by the enzyme sensitivity to temperature variations [90].

NON-INVASIVE DETECTION METHODS

Due to the huge economic losses caused by the new emerging pathogens in agriculture, more sensitive detection methods are needed for diagnostic accuracy and right decision making. The spectroscopic and imaging techniques are non-destructive monitoring methods that have been employed to detect diseases at the early onset of symptoms [91]. They are currently used to create a practical tool for a large-scale disease survey under field conditions [91]. As non-invasive methods, remote sensing allows monitoring of physiological and anatomical changes rather than isolating the pathogen as with conventional techniques [92]. Many spectroscopic and imaging approaches such as, fluorescence imaging, infrared spectroscopy, and fluorescence spectroscopy have been adopted for detecting plant diseases at different stages of infection [91, 93]. These techniques were also employed to detect asymptomatic plant diseases, which is of crucial importance in the design of successful management programs aiming to restrict disease propagation [94]. These techniques were also applied for monitoring postharvest diseases.

Imaging Methods

Thermography

Previous research studies have reported that infection by phytopathogens impact water loss in plants and change the surface temperature of the host tissues [95]. Based on these observations, a technique called thermography was designed to detect the variation in surface temperature of plant leaves and canopies, which aid in the early detection of biotic diseases [96]. The general concept of this technique is based on the record of infrared radiation on the surface of plant leaves by thermographic cameras. The output is a false color image, where each pixel contains the temperature value of the measured object. Thermography is also a promising tool to monitor the heterogeneity in the infection of soil-borne pathogens [97]. However, the application of thermography for disease survey is limited due to its high sensitivity to the change of environmental conditions during measurements. Additionally, thermographic detection lacks the specificity towards diseases, and therefore cannot be used to identify the type of infection or distinguish between diseases that produce similar thermographic patterns. However, not only foliar pathogens can induce local and well-defined physiological changes, but also disorders caused by soil-borne pathogens usually impact the transpiration rate and the water flow of the entire plant [98]. Imaging of leaf and canopy temperature by thermography has been widely employed in plant phenotyping, mainly to characterize susceptibility to drought stress [99]. In fact, leaf temperature is negatively correlated with transpiration and stomatal conductance, which is deeply regulated by plant hosts as a general mechanism of defense on abiotic stress, but also against pathogen attacks [100]. This technique has been employed in the study of plant diseases incited by virus, bacteria, and fungi. For instance, thermography was used to assess cucumber diseased plants with *Pseudoperonospora cubensis* or apple trees infected by *Venturia inaequalis* [101].

Fluorescence Imaging

This technique, also called chlorophyll fluorescence imaging, is a well-established tool for a reliable assessment of pathogen infection and its effect on the leaves of many host plants species [102]. This technique is applicable for entire plants, detached leaves and leaf fragments taken from infected plant tissues [102, 103]. Combining fluorescence imaging with image analysis techniques was useful for discrimination of fungal infections. However, the main disadvantage of this technique is the difficulty of the protocol adopted for plant preparation, which is difficult to apply in normal agricultural greenhouses or field conditions. Therefore, research has been directed at extracting fluorescence parameters from sun-induced reflectance in the field, which would have potential for plant disease assessment at the canopy or field level [104].

In this method, the chlorophyll fluorescence is measured on the leaves as a function of the incident light and the change in fluorescence parameters can be used to analyze pathogen infections, based on changes in the photosynthetic apparatus and photosynthetic electron transport reactions (Fig. **5**) [107]. Using this technique, temporal and spatial variations of chlorophyll fluorescence were analyzed for precise detection of leaf rust and powdery mildew infections in wheat leaves at 470 nm [108]. Although fluorescence measurement provides sensitive detection of abnormalities in photosynthesis, the practical application of this technique in a field setting is limited [109].

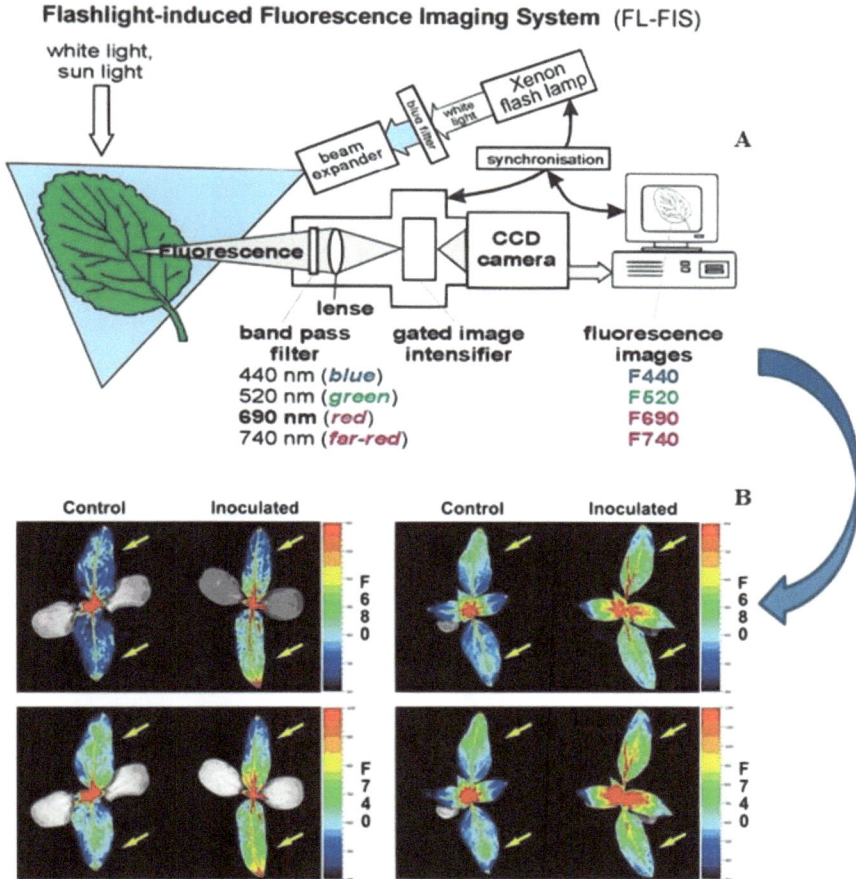

Fig. (5). An outline representing a fluorescence imaging system to measure the blue, green, red and far-red fluorescence signals of whole leaves. **(A)**: The emitted Chlorophyll fluorescence is recorded using a charged-coupled device (CCD) camera and analyzed by computer software to generate pseudo-color images. **(B)**: Red fluorescence (F680) and far-red fluorescence (F740) images of sunflower plants inoculated with *Orobanche cumana* and control plants at different time points [105, 106].

Hyperspectral Imaging

Among the different sensor types, hyperspectral sensors have significant advantages for detecting plant diseases and host–pathogen interaction. Hyperspectral sensors are specific and able to identify the disease and its causal pathogen, which is not allowed by thermography and fluorescence imaging [92].

The main challenge in hyperspectral imaging applied for detecting plant disease, is the finding of the right spectral bands for the target disease (Fig. **6**). In fact, infected plants develop specific physiological changes, which allow the generation of specific disease signature that can be detected by hyperspectral imaging [110]. Therefore, it is possible to detect disease outbreaks and predict the pathogen dynamics by hyperspectral imaging [112]. This allows early detection of specific plant diseases in early stages or even before they are detectable by visual inspection [112]. This advantage facilitates the design and application of the right disease management program. Several research studies using hyperspectral imaging proved that high spatial resolution is critical to avoid mixed spectral signals [92]. Hyperspectral imaging was applied to detect many fungal diseases in the field [110] and in the laboratory [92]. The importance of the spatial resolution was shown by small-scale assays for detecting Cercospora leaf spot, rust and powdery mildew on sugar beet [92]. Also, hyperspectral imaging was applied to specifically detect Fusarium head blight infecting wheat using sensor with a wavelength range of 425–860 nm [113]. Hyperspectral imaging was also applied for detecting Cercospora leaf spot disease in sugar beet [114].

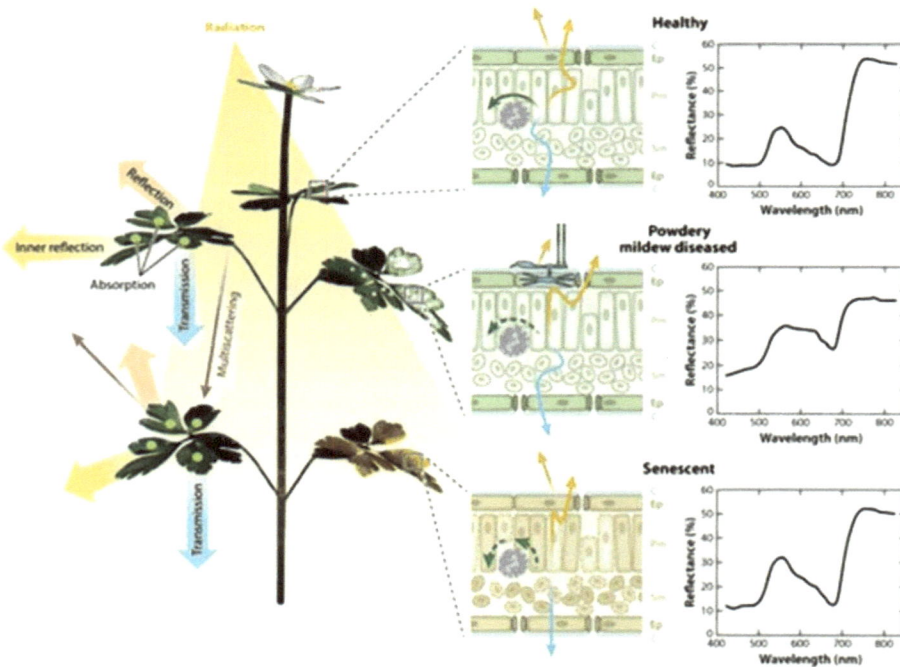

Fig. (6). Comparison between the spectral reflectance of the healthy, senescent and infected leaves. The proportion and intensity of light that is reflected relies on the physiological status of the leaf, which is converted to specific signatures that can be assigned to the health status or disease infection [115].

DETECTION OF PLANT PATHOGENS USING BIOSENSORS

A biosensor is a compact analytical device incorporating a sensitive recognition molecule in direct contact with a suitable physiochemical transducer to assess biochemical variations occurring at the sensor surface. Subsequently, the biological recognition response is converted into an electrical signal, which is further processed and displayed. Pathogen biosensing approaches are optimized using different types of receptors such as DNA\RNA probes, enzymes, bacteriophages and antibodies (Fig. 7) [29, 116]

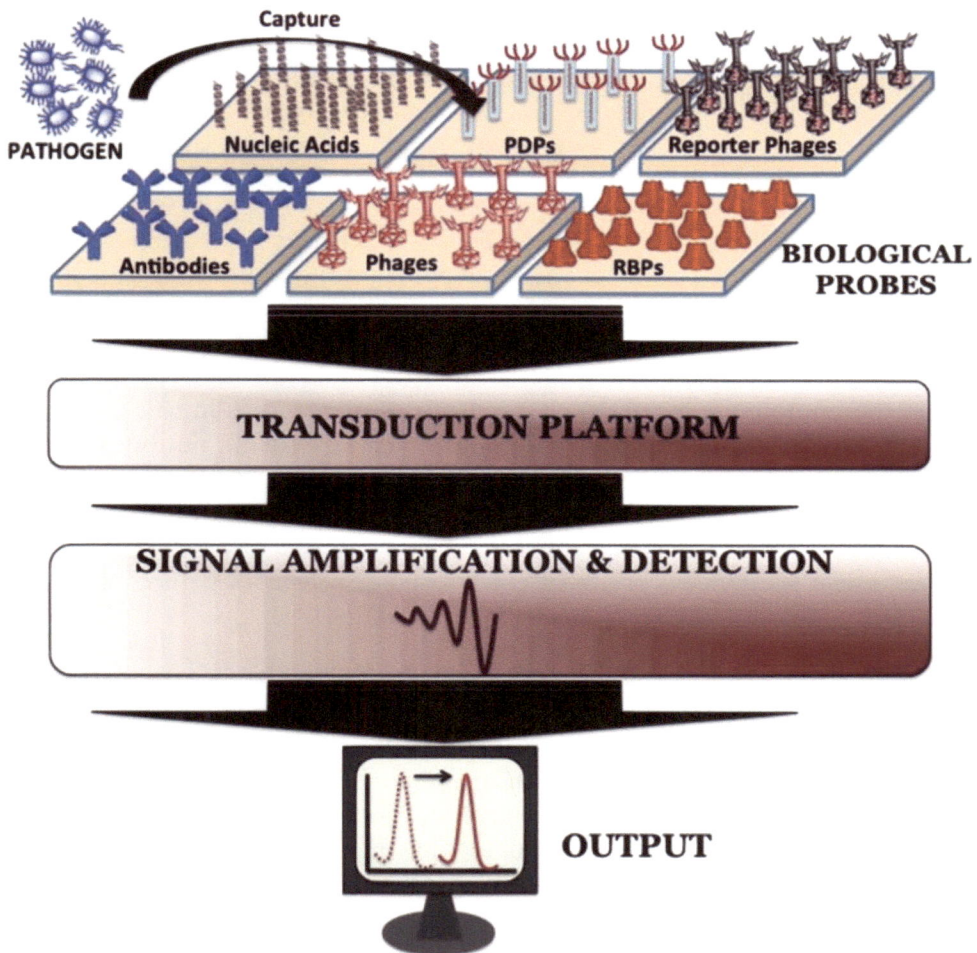

Fig. (7). Pathogen identification approaches using different biological recognition probes including antibodies, DNA/RNA probes, phage display and receptor binding proteins [110].

Antibodies are the most common recognition materials for bio-sensing because of their versatility. Antibodies are sensitive and provide rapid detection and recognition of the target pathogen [117]. Most antibody-based biosensors often use electrochemical transducers, which converts the specific interaction between antibody and antigen to a signal that can be analyzed [118]. Recently, more types of transducers have been used to develop antibody-based sensors, such as surface plasmon resonance, quartz crystal microbalance and cantilever-based sensors. Recently, several research studies have proved the reliability of antibody-based biosensors for detecting viral, fungal and bacterial plant pathogens [118, 119]. Despite their wide application, they still have some limits since specific binding of antibody with a particular antigen, is influenced by environmental conditions such as pH and temperature. Interestingly, antibodies are sensitive for storage conditions and likely subjected to deterioration over time. In addition, immobilizing large bacteria and fungi, whose sizes exceed the transducer range, might restrict the detection [116].

Since single strand DNA (ssDNA) exhibits high affinity to its complementary strand, DNA can be used as a sensitive recognition molecule to develop DNA-based biosensor. This technique allows early detection of pathogens before the onset of visual symptoms. The most commonly adopted DNA probe is ssDNA on electrodes with electro-active signals to measure hybridization between probe DNA and the complementary DNA present in tested samples [120]. DNA based biosensors exhibit higher sensitivity compared to those using antibodies, which facilitate the detection of pathogens with low titer inside host tissues. However, they have some limits related to the selection and synthesis of specific DNA probes. In addition, the sensitivity and specificity of the technique are affected when short DNA sequences are selected and used as probe for pathogen detection [121].

Adopting enzymes as biorecognition molecule can provide specific detection of the target pathogen [122]. For plant disease diagnosis, enzymatic based biosensors could be adopted by targeting Volatile Organic Compounds (VOCs) being released by the infected plant [123]. This new technology assumes that each disease has its own VOCs signature profile. Thus, by defining the type and concentration of VOCs being released by the diseased plant, we can assume whether a plant is infected and the type of disease it has [123]. In fact, many of the VOCs generated by infected crops are alcohols and aldehydes such as cis--hexen-1-ol and trans-2-hexanal, that can be catalyzed by alcohol dehydrogenase enzymes, which in their turn could serve as biosensors for detecting VOCs specific to the infection [29]. The common methods used for analyzing the VOCs signature are gas chromatography (GC)-based and electronic nose system-based techniques. In fact, specific volatile biomarker may be used to detect plant diseases as reported for the potato soft rot caused by *Erwinia carotovora* [124].

Researchers at North Carolina State University have invented portable technology that allows farmers and pathologists to detect and identify plant diseases in the field. The device, which is plugged into a smartphone, works by sampling the VOCs released by the leaves taken from infected plants [6].

CONCLUSIONS

In this chapter, a wide variety of advanced detection techniques adopted for plant pathogens diagnosis have been reviewed. Although visual inspection methods and deep knowledge of symptoms etiologies are essential for assessing plant diseases, the latest technologies including molecular methods, imaging, and VOCs profiling detection methods made the pathogen identification more convenient, faster and more sensitive. Molecular and serological methods have been often used to detect plant diseases with acceptable specificity, ease and cost. However, due to their inability to provide real-time detection, makes them unsuitable for on-site diagnosis and early warning. New technologies such as fluorescence imaging and biosensors-based techniques brings benefits such as identification of unculturable microorganisms, early detection of infection even in symptomless plants and the ability to perform large scale diagnosis. The specificity and sensitivity of the biosensors could be improved by using enzymes, antibodies, DNA and bacteriophage.

CONSENT FOR PUBLICATION

Not applicable.

CONFLICT OF INTEREST

The author(s) confirms that there is no conflict of interest.

ACKNOWLEDGEMENTS

Declared none.

REFERENCES

[1]   Savary S, Ficke A, Aubertot J, Hollier C. Crop losses due to diseases and their implications for global food production losses and food security. Food Secur 2012; 4: 519-37.
       [http://dx.doi.org/10.1007/s12571-012-0200-5]

[2]   Oerke EC. Crop losses to pests. J Agric Sci 2006; 144: 31-43.
       [http://dx.doi.org/10.1017/S0021859605005708]

[3]   Kumar P, Akhtar J, Kandan A, Kumar S, Batra R, Dubey SC. Advance detection techniques of phytopathogenic fungi: Current trends and future perspectives.Current Trends in Plant Disease Diagnostics and Management Practices Fungal Biology. Cham: Springer 2016; pp. 265-98.

[http://dx.doi.org/10.1007/978-3-319-27312-9_12]

[4]   Tan DHS, Sigler L, Gibas CFC, Fong IW. Disseminated fungal infection in a renal transplant recipient involving *Macrophomina phaseolina* and *Scytalidium dimidiatum*: case report and review of taxonomic changes among medically important members of the Botryosphaeriaceae. Med Mycol 2008; 46(3): 285-92.
[http://dx.doi.org/10.1080/13693780701759658] [PMID: 18404556]

[5]   Aslam S, Tahir A, Aslam MF, Alam MW, Shedayi AA, Sadia S. Recent advances in molecular techniques for the identification of phytopathogenic fungi – a mini review. J Plant Interact 2017; 12(1): 493-504.
[http://dx.doi.org/10.1080/17429145.2017.1397205]

[6]   Li Z, Paul R, Ba Tis T, *et al.* Non-invasive plant disease diagnostics enabled by smartphone-based fingerprinting of leaf volatiles. Nat Plants 2019; 5(8): 856-66.
[http://dx.doi.org/10.1038/s41477-019-0476-y] [PMID: 31358961]

[7]   Harrison JG, Lowe R, Duncan JM. Use of ELISA for assessing major gene resistance of potato leaves to *Phytophthora infestans*. Plant Pathol 1991; 40: 431-5.
[http://dx.doi.org/10.1111/j.1365-3059.1991.tb02401.x]

[8]   Skottrup P, Nicolaisen M, Justesen AF. Rapid determination of *Phytophthora infestans* sporangia using a surface plasmon resonance immunosensor. J Microbiol Methods 2007; 68(3): 507-15.
[http://dx.doi.org/10.1016/j.mimet.2006.10.011] [PMID: 17157943]

[9]   Lees AK, Sullivan L, Lynott JS, Cullen DW. Development of a quantitative real-time PCR assay for *Phytophthora infestans* and its applicability to leaf, tuber and soil samples. Plant Pathol 2012; 61: 867-76.
[http://dx.doi.org/10.1111/j.1365-3059.2011.02574.x]

[10]   Hansen ZR, Knaus BJ, Tabima JF, *et al.* Loop-mediated isothermal amplification for detection of the tomato and potato late blight pathogen, *Phytophthora infestans*. J Appl Microbiol 2016; 120(4): 1010-20.
[http://dx.doi.org/10.1111/jam.13079] [PMID: 26820117]

[11]   Clark MF. Immunosorbent assays in plant pathology. Annu Rev Phytopathol 1981; 19: 83-106.
[http://dx.doi.org/10.1146/annurev.py.19.090181.000503]

[12]   Bailey AM, Mitchell DJ, Manjunath KL, Nolasco G, Niblett CL. Identification to the species level of the plant pathogens *Phytophthora* and *Pythium* by using unique sequences of the ITS1 region of ribosomal DNA as capture probes for PCR ELISA. FEMS Microbiol Lett 2002; 207(2): 153-8.
[http://dx.doi.org/10.1111/j.1574-6968.2002.tb11044.x] [PMID: 11958933]

[13]   Chandra NS, Wulff EG, Udayashankar AC, *et al.* Prospects of molecular markers in Fusarium species diversity. Appl Microbiol Biotechnol 2011; 90(5): 1625-39.
[http://dx.doi.org/10.1007/s00253-011-3209-3] [PMID: 21494869]

[14]   Waliullah S, Hudson O, Oliver JE, Brannen PM, Ji P, Ali ME. Comparative analysis of different molecular and serological methods for detection of *Xylella fastidiosa* in blueberry. PLoS One 2019; 14(9)e0221903
[http://dx.doi.org/10.1371/journal.pone.0221903] [PMID: 31479482]

[15]   Atallah ZK, Bae J, Jansky SH, Rouse DI, Stevenson WR. Multiplex real-time quantitative PCR to detect and quantify *Verticillium dahliae* colonization in potato lines that differ in response to *Verticillium* wilt. Phytopathology 2007; 97(7): 865-72.
[http://dx.doi.org/10.1094/PHYTO-97-7-0865] [PMID: 18943936]

[16]   Babu BK, Mesapogu S, Sharma A, Somasani SR, Arora DK. Quantitative real-time PCR assay for rapid detection of plant and human pathogenic *Macrophomina phaseolina* from field and environmental samples. Mycologia 2011; 103(3): 466-73.
[http://dx.doi.org/10.3852/10-181] [PMID: 21186328]

[17]   Lau HY, Botella JR. Advanced DNA-based point-of-care diagnostic methods for plant diseases detection. Front Plant Sci 2017; 8: 2016.
[http://dx.doi.org/10.3389/fpls.2017.02016] [PMID: 29375588]

[18]   Martin KJ, Rygiewicz PT. Fungal-specific PCR primers developed for analysis of the ITS region of environmental DNA extracts. BMC Microbiol 2005; 5: 28.
[http://dx.doi.org/10.1186/1471-2180-5-28] [PMID: 15904497]

[19]   Durai M, Dubey SC, Tripathi A. Genetic diversity analysis and development of SCAR marker for detection of Indian populations of *Fusarium oxysporum* f. sp. ciceris causing chickpea wilt. Folia Microbiol (Praha) 2012; 57(3): 229-35.
[http://dx.doi.org/10.1007/s12223-012-0118-5] [PMID: 22528298]

[20]   Ganeshamoorthi P, Dubey SC. Phylogeny analysis of Indian strains of *Rhizoctonia solani* isolated from chickpea and development of sequence characterized amplifi ed region (SCAR) marker for detection of the pathogen. Afr J Microbiol Res 2013; 7: 5516-25.
[http://dx.doi.org/10.5897/AJMR2013.5769]

[21]   Luchi N, Ioos R, Santini A. Fast and reliable molecular methods to detect fungal pathogens in woody plants. Appl Microbiol Biotechnol 2020; 104(6): 2453-68.
[http://dx.doi.org/10.1007/s00253-020-10395-4] [PMID: 32006049]

[22]   Okubara PA, Schroeder KL, Paulitz TC. Real-time polymerase chain reaction: Applications to studies on soil borne pathogens. Can J Plant Pathol 2005; 27: 300-13.
[http://dx.doi.org/10.1080/07060660509507229]

[23]   Mahuku GS, Goodwin PH. Presence of *Xanthomonas fragariae* in symptomless strawberry crowns in Ontario detected using a nested polymerase chain reaction (PCR). Can J Plant Pathol 1997; 19(4): 366-70.
[http://dx.doi.org/10.1080/07060669709501061]

[24]   Bilodeau GJ, Koike ST, Uribe P, Martin FN. Development of an assay for rapid detection and quantification of *Verticillium dahliae* in soil. Phytopathology 2012; 102(3): 331-43.
[http://dx.doi.org/10.1094/PHYTO-05-11-0130] [PMID: 22066673]

[25]   Ahrberg CD, Ilic BR, Manz A, Neužil P. Handheld real-time PCR device. Lab Chip 2016; 16(3): 586-92.
[http://dx.doi.org/10.1039/C5LC01415H] [PMID: 26753557]

[26]   Notomi T, Okayama H, Masubuchi H, *et al.* Loop-mediated isothermal amplification of DNA. Nucleic Acids Res 2000; 28(12)E63
[http://dx.doi.org/10.1093/nar/28.12.e63] [PMID: 10871386]

[27]   Calderón R, Navas-Cortés JA, Lucena C, Zarco-Tejada PJ. High-resolution airborne hyperspectral and thermal imagery for early detection of Verticillium wilt of olive using fluorescence, temperature and narrow-band spectral indices. Remote Sens Environ 2013; 139: 231-45.
[http://dx.doi.org/10.1016/j.rse.2013.07.031]

[28]   Mahlein AK. Plant Disease Detection by Imaging Sensors - Parallels and Specific Demands for Precision Agriculture and Plant Phenotyping. Plant Dis 2016; 100(2): 241-51.
[http://dx.doi.org/10.1094/PDIS-03-15-0340-FE] [PMID: 30694129]

[29]   Fang Y, Ramasamy RP. Current and Prospective Methods for Plant Disease Detection. Biosensors (Basel) 2015; 5(3): 537-61.
[http://dx.doi.org/10.3390/bios5030537] [PMID: 26287253]

[30]   Cao X, Ye Y, Liu S. Gold nanoparticle-based signal amplification for biosensing. Anal Biochem 2011; 417(1): 1-16.
[http://dx.doi.org/10.1016/j.ab.2011.05.027] [PMID: 21703222]

[31]   Kuila T, Bose S, Khanra P, Mishra AK, Kim NH, Lee JH. Recent advances in graphene-based biosensors. Biosens Bioelectron 2011; 26(12): 4637-48.

[http://dx.doi.org/10.1016/j.bios.2011.05.039] [PMID: 21683572]

[32]   Mandler D, Kraus-Ophir S. Self-assembled monolayers (SAMs) for electrochemical sensing. J Solid State Electrochem 2011; 15: 1535-58.
[http://dx.doi.org/10.1007/s10008-011-1493-6]

[33]   Wang T, Li P, Zhang Q, *et al*. Determination of *Aspergillus* pathogens in agricultural products by a specific nanobody-polyclonal antibody sandwich ELISA. Sci Rep 2017; 7(1): 4348.
[http://dx.doi.org/10.1038/s41598-017-04195-6] [PMID: 28659622]

[34]   Widershain GY. The ELISA Guidebook. Biochemistry (Mosc) 2009; 74: 1058.
[http://dx.doi.org/10.1134/S000629790909017X]

[35]   Cimmino A, Iannaccone M, Petriccione M, *et al*. An ELISA method to identify the phytotoxic *Pseudomonas syringae* pv. *actinidiae* exopolysaccharides: A tool for rapid immunochemical detection of kiwifruit bacterial canker. Phytochem Lett 2017; 19: 136-40.
[http://dx.doi.org/10.1016/j.phytol.2016.12.027]

[36]   Rajeshwari N, Shylaja M, Krishnappa M, Shetty HS, Mortensen CN, Mathur SB. Development of ELISA for the detection of *Ralstonia solanacearum* in tomato: its application in seed health testing. World J Microbiol Biotechnol 1998; 14: 697-704.
[http://dx.doi.org/10.1023/A:1008892400077]

[37]   Kokosková B, Mráz I, Hyblová J. Comparison of specificity and sensitivity of immunochemical and molecular techniques for reliable detection of *Erwinia amylovora*. Folia Microbiol (Praha) 2007; 52(2): 175-82.
[http://dx.doi.org/10.1007/BF02932156] [PMID: 17575916]

[38]   Tsror L, Erlich O, Hazanovsky M, Ben Daniel B, Zig U, Lebiush S. Detection of Dickeya spp. latent infection in potato seed tubers using PCR or ELISA and correlation with disease incidence in commercial field crops under hot-climate conditions. Plant Pathol 2012; 61: 161-8.
[http://dx.doi.org/10.1111/j.1365-3059.2011.02492.x]

[39]   Nader AA, Said IB, Hosny AY, Ahmed EK. Development of polyclonal rabbit serum-based ELISA for Detection of *Pectobacterium carotovorum* subsp. *carotovorum* and its Specificity against other Causing Soft Rot Bacteria. Asian J Plant Pathol 2015; 9: 135-41.
[http://dx.doi.org/10.3923/ajppaj.2015.135.141]

[40]   Holzlöhner P, Hanack K. Generation of murine monoclonal antibodies by hybridoma technology. J Vis Exp 2017; 119(119)e54832
[http://dx.doi.org/10.3791/54832] [PMID: 28117810]

[41]   Scala V, Pucci N, Loreti S. The diagnosis of plant pathogenic bacteria: A state of art. Front Biosci (Elite Ed) 2018; 10: 449-60.
[http://dx.doi.org/10.2741/e832] [PMID: 29293468]

[42]   Song G, Wu JY, Xie Y, *et al*. Monoclonal antibody-based serological assays for detection of Potato virus S in potato plants. J Zhejiang Univ Sci B 2017; 18(12): 1075-82.
[http://dx.doi.org/10.1631/jzus.B1600561] [PMID: 29204987]

[43]   Janse JD, Kokoskova B. Indirect immunofluorescence microscopy for the detection and identification of plant pathogenic bacteria (in particular for *Ralstonia solanacearum*. In: Burns R, Ed. Plant Pathology Methods in Molecular Biology (Methods and Protocols). Totowa, NJ: Humana Press 2009; pp. 89-99.
[http://dx.doi.org/10.1007/978-1-59745-062-1_8]

[44]   Vittal R, Haudenshield JS, Hartman GL. A multiplexed immunofluorescence method identifies Phakopsora pachyrhizi Urediniospores and determines their viability. Phytopathology 2012; 102(12): 1143-52.
[http://dx.doi.org/10.1094/PHYTO-02-12-0040-R] [PMID: 22894915]

[45]   Przewodowski W, Przewodowska A. Development of a sensitive and specific polyclonal antibody for

serological detection of *Clavibacter michiganensis* subsp. *sepedonicus.* PLoS One 2017; 12(1)e0169785
[http://dx.doi.org/10.1371/journal.pone.0169785] [PMID: 28068400]

[46] López MM, Bertolini E, Olmos A, *et al.* Innovative tools for detection of plant pathogenic viruses and bacteria. Int Microbiol 2003; 6(4): 233-43.
[http://dx.doi.org/10.1007/s10123-003-0143-y] [PMID: 13680391]

[47] Gu L, Bai Z, Jin B, *et al.* Assessing the impact of fungicide enostroburin application on bacterial community in wheat phyllosphere. J Environ Sci (China) 2010; 22(1): 134-41.
[http://dx.doi.org/10.1016/S1001-0742(09)60084-X] [PMID: 20397397]

[48] Aloisio M, Morelli M, Elicio V, *et al.* Detection of four regulated grapevine viruses in a qualitative, single tube real-time PCR with melting curve analysis. J Virol Methods 2018; 257: 42-7.
[http://dx.doi.org/10.1016/j.jviromet.2018.04.008] [PMID: 29654789]

[49] Chen S, Cao Y, Li T, Wu X. Simultaneous detection of three wheat pathogenic fungal species by multiplex PCR. Phytoparasitica 2014; 43: 449-60.
[http://dx.doi.org/10.1007/s12600-014-0442-1]

[50] Hyun JW, Hwang RY, Jung KE. Development of Multiplex PCR for Simultaneous Detection of Citrus Viruses and the incidence of citrus viral diseases in late-maturity citrus trees in Jeju Island. Plant Pathol J 2017; 33(3): 307-17.
[http://dx.doi.org/10.5423/PPJ.OA.10.2016.0207] [PMID: 28592949]

[51] Lefeuvre P, Hoareau M, Delatte H, Reynaud B, Lett JM. A multiplex PCR method discriminating between the TYLCV and TYLCV-Mld clades of *tomato yellow leaf curl virus.* J Virol Methods 2007; 144(1-2): 165-8.
[http://dx.doi.org/10.1016/j.jviromet.2007.03.020] [PMID: 17485124]

[52] Zeng QY, Hansson P, Wang XR. Specific and sensitive detection of the conifer pathogen *Gremmeniella abietina* by nested PCR. BMC Microbiol 2005; 5: 65.
[http://dx.doi.org/10.1186/1471-2180-5-65] [PMID: 16280082]

[53] Yang X, Hameed U, Zhang AF, *et al.* Development of a nested-PCR assay for the rapid detection of *Pilidiella granati* in pomegranate fruit. Sci Rep 2017; 7: 40954.
[http://dx.doi.org/10.1038/srep40954] [PMID: 28106107]

[54] Pérez-Hernández O, Nam MH, Gleason ML, Kim HG. Development of a nested polymerase chain reaction assay for detection of *Colletotrichum acutatum* on symptomless strawberry leaves. Plant Dis 2008; 92(12): 1655-61.
[http://dx.doi.org/10.1094/PDIS-92-12-1655] [PMID: 30764297]

[55] Glen M, Smith AH, Langrell SR, Mohammed CL. Development of nested polymerase chain reaction detection of mycosphaerella spp. and its application to the study of leaf disease in eucalyptus plantations. Phytopathology 2007; 97(2): 132-44.
[http://dx.doi.org/10.1094/PHYTO-97-2-0132] [PMID: 18944368]

[56] Khasa E, Gopala A, Taloh A, Prabha T, Pria M, Rao GP. Molecular characterization of phytoplasmas of 'Clover proliferation' group associated with three ornamental plant species in India. 3 Biotech 2016; 6: 237.

[57] Wei S, Sun Y, Xi G, Zhang H, Xiao M, Yin R. Development of a single-tube nested PCR-lateral flow biosensor assay for rapid and accurate detection of *Alternaria panax* Whetz. PLoS One 2018; 13(11)e0206462
[http://dx.doi.org/10.1371/journal.pone.0206462] [PMID: 30408825]

[58] Bernal-Galeano V, Ochoa JC, Trujillo C, Rache L, Bernal A, Lopez CA. Development of a multiplex nested PCR method for detection of *Xanthomonas axonopodis* pv. *manihotis* in Cassava. Trop Plant Pathol 2018; 43: 341-50.
[http://dx.doi.org/10.1007/s40858-018-0214-4]

[59]   Adriko J, Aritua V, Mortensen CN, Tushemereirwe WK, Kubiriba J, Lund OS. Multiplex PCR for specific and robust detection of *Xanthomonas campestris* pv. *musacearum* in pure culture and infected plant material. Plant Pathol 2012; 61: 489-97.
        [http://dx.doi.org/10.1111/j.1365-3059.2011.02534.x]

[60]   Collado-Romero M, Berbegal M, Jiménez-Díaz RM, Armengol J, Mercado-Blanco J. A PCR-based 'molecular tool box' for *in planta* differential detection of *Verticillium dahliae* vegetative compatibility groups infecting artichoke. Plant Pathol 2009; 58: 515-26.
        [http://dx.doi.org/10.1111/j.1365-3059.2008.01981.x]

[61]   Robène-Soustrade I, Legrand D, Gagnevin L, Chiroleu F, Laurent A, Pruvost O. Multiplex nested PCR for detection of Xanthomonas axonopodis pv. allii from onion seeds. Appl Environ Microbiol 2010; 76(9): 2697-703.
        [http://dx.doi.org/10.1128/AEM.02697-09] [PMID: 20208024]

[62]   Koolivand D, Sokhandan-Bashir N, Behjatnia SAB, Joozani RAJ. Detection of *Grapevine fanleaf virus* by immunocapture reverse transcription-polymerase chain reaction (IC-RT-PCR) with recombinant antibody. Arch Phytopathol Pflanzenschutz 2014; 47(17): 2070-7.
        [http://dx.doi.org/10.1080/03235408.2013.868697]

[63]   Mulholland V. Immunocapture-PCR for plant virus detection. In: Burns R, Ed. Plant Pathology, Plant Pathology Methods in Molecular Biology (Methods and Protocols). Totowa, NJ: Humana Press 2009; pp. 183-92.

[64]   Gharbi Y, Barkallah M, Bouazizi E, Cheffi M, Gdoura R, Triki MA. Differential fungal colonization and physiological defense responses of new olive cultivars infected by the necrotrophic fungus *Verticillium dahliae*. Acta Physiol Plant 2016; 38: 242.
        [http://dx.doi.org/10.1007/s11738-016-2261-0]

[65]   Gharbi Y, Barkallah M, Bouazizi E, *et al.* Development and validation of a new real-time assay for the quantification of *Verticillium dahliae* in the soil: a comparison with conventional soil plating. Mycol Prog 2016; 15: 54.
        [http://dx.doi.org/10.1007/s11557-016-1196-6]

[66]   Dubey SC, Priyanka K, Upadhyay BK. Development of molecular markers and probes based on TEF-1α, β-tubulin and ITS gene sequences for quantitative detection of *Fusarium oxysporum* f. sp. ciceris by using real-time PCR. Phytoparasitica 2014; 42: 355-66.
        [http://dx.doi.org/10.1007/s12600-013-0369-y]

[67]   Gudnason H, Dufva M, Bang DD, Wolff A. Comparison of multiple DNA dyes for real-time PCR: effects of dye concentration and sequence composition on DNA amplification and melting temperature. Nucleic Acids Res 2007; 35(19)e127
        [http://dx.doi.org/10.1093/nar/gkm671] [PMID: 17897966]

[68]   Schena L, Nigro F, Ippolito A, Gallitelli D. Real-time quantitative PCR: A new technology to detect and study phytopathogenic and antagonistic fungi. Eur J Plant Pathol 2014; 110: 893-908.
        [http://dx.doi.org/10.1007/s10658-004-4842-9]

[69]   Liu M, McCabe E, Chapados JT, *et al.* Detection and identification of selected cereal rust pathogens by TaqMan® real-time PCR. Can J Plant Pathol 2015; 37: 92-105.
        [http://dx.doi.org/10.1080/07060661.2014.999123]

[70]   Gharbi Y, Barkallah M, Bouazizi E, Hibar K, Gdoura R, Triki MA. Lignification, phenols accumulation, induction of PR proteins and antioxidant-related enzymes are key factors in the resistance of *Olea europaea* to Verticillium wilt of olive. Acta Physiol Plant 2017; 39: 43.
        [http://dx.doi.org/10.1007/s11738-016-2343-z]

[71]   Cregeen S, Radisek S, Mandelc S, *et al.* Different gene expressions of resistant and susceptible hop cultivars in response to infection with a highly aggressive strain of *Verticillium albo-atrum*. Plant Mol Biol Report 2015; 33(3): 689-704.
        [http://dx.doi.org/10.1007/s11105-014-0767-4] [PMID: 25999664]

[72] Singh N, Kapoor R. Quick and accurate detection of *Fusarium oxysporum* f. sp. *carthami* in host tissue and soil using conventional and real-time PCR assay. World J Microbiol Biotechnol 2018; 34(12): 175.
[http://dx.doi.org/10.1007/s11274-018-2556-y] [PMID: 30446834]

[73] Strayer AL, Jeyaprakash A, Minsavage GV, *et al.* A multiplex real-time PCR assay differentiates four *Xanthomonas* species associated with bacterial spot of tomato. Plant Dis 2016; 100(8): 1660-8.
[http://dx.doi.org/10.1094/PDIS-09-15-1085-RE] [PMID: 30686244]

[74] Nanayakkara UN, Singh M, Al-Mughrabi KI, Peters RD. Detection of phytophthora erythroseptica in above-ground potato tissues, progeny tubers, stolons and crop debris using PCR techniques. Am J Potato Res 2009; 86(3): 239-45.
[http://dx.doi.org/10.1007/s12230-009-9077-z]

[75] Schroeder KL, Martin FN, de Cock AWAM, *et al.* Molecular detection and quantification of *Pythium* species: evolving taxonomy, new tools, and challenges. Plant Dis 2013; 97(1): 4-20.
[http://dx.doi.org/10.1094/PDIS-03-12-0243-FE] [PMID: 30722255]

[76] Holeva MC, Morán F, Scuderi G, González A, López MM, Llop P. Development of a real-time PCR method for the specific detection of the novel pear pathogen *Erwinia uzenensis*. PLoS One 2019; 14(7)e0219487
[http://dx.doi.org/10.1371/journal.pone.0219487] [PMID: 31291321]

[77] Harper SJ, Ward LI, Clover GRG. Development of LAMP and real-time PCR methods for the rapid detection of *Xylella fastidiosa* for quarantine and field applications. Phytopathology 2010; 100(12): 1282-8.
[http://dx.doi.org/10.1094/PHYTO-06-10-0168] [PMID: 20731533]

[78] Hinze M, Köhl L, Kunz S, *et al.* Real☐time PCR detection of *Erwinia amylovora* on blossoms correlates with subsequent fire blight incidence. Plant Pathol 2016; 65: 462-9.
[http://dx.doi.org/10.1111/ppa.12429]

[79] Duan YB, Ge CY, Zhang XK, Wang JX, Zhou MG. Development and evaluation of a novel and rapid detection assay for *Botrytis cinerea* based on loop-mediated isothermal amplification. PLoS One 2014; 9(10)e111094 b
[http://dx.doi.org/10.1371/journal.pone.0111094] [PMID: 25329402]

[80] Aglietti C, Luchi N, Pepori AL, *et al.* Real-time loop-mediated isothermal amplification: an early-warning tool for quarantine plant pathogen detection. AMB Express 2019; 9(1): 50.
[http://dx.doi.org/10.1186/s13568-019-0774-9] [PMID: 31016406]

[81] Harrison C, Tomlinson J, Ostoja-Starzewska S, Boonham N. Evaluation and validation of a loop-mediated isothermal amplification test kit for detection of *Hymenoscyphus fraxineus*. Eur J Plant Pathol 2017; 149: 253-9.
[http://dx.doi.org/10.1007/s10658-017-1179-8]

[82] Wong B, Leal I, Feau N, Dale A, Uzunovic A, Hamelin RC. Molecular assays to detect the presence and viability of *Phytophthora ramorum* and *Grosmannia clavigera*. PLoS One 2020; 15(2)e0221742
[http://dx.doi.org/10.1371/journal.pone.0221742] [PMID: 32023247]

[83] Guo JR, Schnieder F, Beyer M, Verreet J-A. Rapid detection of *Mycosphaerella graminicola* in wheat using reverse transcription-PCR assay. J Phytopathol 2005; 153: 674-9.
[http://dx.doi.org/10.1111/j.1439-0434.2005.01035.x]

[84] Abdullah AS, Turo C, Moffat CS, *et al.* Real-Time PCR for diagnosing and quantifying co-infection by two globally distributed fungal pathogens of wheat. Front Plant Sci 2018; 9: 1086.
[http://dx.doi.org/10.3389/fpls.2018.01086] [PMID: 30140271]

[85] Brown NA, Bass C, Baldwin TK, *et al.* Characterisation of the *Fusarium graminearum*-wheat floral interaction. J Pathogens 2011; 2011626345
[http://dx.doi.org/10.4061/2011/626345] [PMID: 22567335]

[86] He Y, Gao F, Shen J, *et al.* A multiplex RT-PCR method for the simultaneous detection of *Narcissus yellow stripe virus, Narcissus latent virus* and *Narcissus mosaic virus*. Can J Plant Pathol 2019; 41(1): 115-23.
[http://dx.doi.org/10.1080/07060661.2018.1513074]

[87] Gan Y, Zhou Z, An L, Bao S, Forde BG. A Comparison cetween Northern blotting and quantitative real-time PCR as a means of detecting the nutritional regulation of genes expressed in roots of *Arabidopsis thaliana*. Agric Sci China 2011; 10: 335-42.
[http://dx.doi.org/10.1016/S1671-2927(11)60012-6]

[88] Kwon SJ, Seo JK, Rao ALN. Detection and quantification of viral and satellite RNAs in plant hosts. In: Jin H, Gassmann W, Eds. RNA Abundance Analysis Methods in Molecular Biology (Methods and Protocols). Totowa, NJ: Humana Press 2012; pp. 131-41.
[http://dx.doi.org/10.1007/978-1-61779-839-9_10]

[89] Fakruddin M, Mazumdar RM, Chowdhury A, Mannan K. Nucleic acid sequence-based amplification (NASBA)-prospects and applications. Int J Life Sci Pharma Res 2012; 2: 106-7.

[90] Chang CC, Chen CC, Wei SC, Lu HH, Liang YH, Lin CW. Diagnostic devices for isothermal nucleic acid amplification. Sensors (Basel) 2012; 12(6): 8319-37.
[http://dx.doi.org/10.3390/s120608319] [PMID: 22969402]

[91] Martinelli F, Scalenghe R, Davino S, Panno S, Scuderi G. Advanced methods of plant disease detection. A review. Agron Sustain Dev 2015; 35(1): 1-25.
[http://dx.doi.org/10.1007/s13593-014-0246-1]

[92] Mahlein AK, Rumpf T, Welke P, *et al.* Development of spectral indices for detecting and identifying plant diseases. Remote Sens Environ 2013; 128: 21-30.
[http://dx.doi.org/10.1016/j.rse.2012.09.019]

[93] Oerke EC, Mahlein AK, Steiner U. Proximal sensing of plant diseases.Detection and Diagnostic of Plant Pathogens, Plant Pathology in the 21st Century.. Dordrecht, the Netherlands: Springer Science and Business Media 2014; pp. 55-68.
[http://dx.doi.org/10.1007/978-94-017-9020-8_4]

[94] Kuska M, Wahabzada M, Leucker M, *et al.* Hyperspectral phenotyping on the microscopic scale: Towards automated characterization of plant-pathogen interactions. Plant Methods 2015; 11: 28.
[http://dx.doi.org/10.1186/s13007-015-0073-7] [PMID: 25937826]

[95] Maes WH, Minchin PEH, Snelgar WP, Steppe K. Early detection of Psa infection in kiwifruit by means of infrared thermography at leaf and orchard scale. Funct Plant Biol 2014; 41(12): 1207-20.
[http://dx.doi.org/10.1071/FP14021] [PMID: 32481070]

[96] Belin E, Rousseau D, Boureau T, Caffier V. Thermography *versus* chlorophyll fluorescence imaging for detection and quantification of apple scab. Comput Electron Agric 2013; 90: 159-63.
[http://dx.doi.org/10.1016/j.compag.2012.09.014]

[97] Hillnhutter C, Mahlein AK, Sikora RA, Oerke EC. Remote sensing to detect plant stress induced by Heterodera schachtii and Rhizoctonia solani in sugar beet fields. Field Crops Res 2011; 122: 70-7.
[http://dx.doi.org/10.1016/j.fcr.2011.02.007]

[98] Granum E, Pérez-Bueno ML, Calderón CE, *et al.* Metabolic responses of avocado plants to stress induced by *Rosellinia necatrix* analysed by fluorescence and thermal imaging. Eur J Plant Pathol 2015; 142: 625-32.
[http://dx.doi.org/10.1007/s10658-015-0640-9]

[99] Pipatsitee P, Eiumnoh A, Praseartkul P, *et al.* Application of infrared thermography to assess cassava physiology under water deficit condition. Plant Prod Sci 2018; 21(4): 398-406.
[http://dx.doi.org/10.1080/1343943X.2018.1530943]

[100] Melotto M, Underwood W, He SY. Role of stomata in plant innate immunity and foliar bacterial diseases. Annu Rev Phytopathol 2008; 46: 101-22.

[http://dx.doi.org/10.1146/annurev.phyto.121107.104959] [PMID: 18422426]

[101] Oerke EC, Frohling P, Steiner U. Thermographic assessment of scab disease on apple leaves. Precis Agric 2011; •••: 699-715.
[http://dx.doi.org/10.1007/s11119-010-9212-3]

[102] Rolfe SA, Scholes JD. Chlorophyll fluorescence imaging of plant-pathogen interactions. Protoplasma 2010; 247(3-4): 163-75.
[http://dx.doi.org/10.1007/s00709-010-0203-z] [PMID: 20814703]

[103] Mahlein A, Oerke E, Steiner U, Dehne HW. Recent advances in sensing plant diseases for precision crop protection. Eur J Plant Pathol 2012; 133: 197-209.
[http://dx.doi.org/10.1007/s10658-011-9878-z]

[104] Rossini M, Nedbal L, Guanter L, *et al.* Red and far red Sun☐induced chlorophyll fluorescence as a measure of plant photosynthesis. Geophys Res Lett 2015; 42: 1632-9.
[http://dx.doi.org/10.1002/2014GL062943]

[105] Lichtenthaler HK, Babani F. 2000.Detection of photosynthetic activity and water stressby imaging the red chlorophyll fluorescence
[http://dx.doi.org/10.1016/S0981-9428(00)01199-2]

[106] Ortiz-Bustos CM, Pérez-Bueno ML, Barón M, Molinero-Ruiz L. Fluorescence imaging in the red and far-red region during growth of sunflower plantlets. diagnosis of the early infection by the parasite orobanche cumana. Front Plant Sci 2016; 7: 884.
[http://dx.doi.org/10.3389/fpls.2016.00884] [PMID: 27446116]

[107] Bürling K, Hunsche M, Noga G. Use of blue-green and chlorophyll fluorescence measurements for differentiation between nitrogen deficiency and pathogen infection in winter wheat. J Plant Physiol 2011; 168(14): 1641-8.
[http://dx.doi.org/10.1016/j.jplph.2011.03.016] [PMID: 21658789]

[108] Kuckenberg J, Tartachnyk I, Noga G. Temporal and spatial changes of chlorophyll fluorescence as a basis for early and precise detection of leaf rust and powdery mildew infections in wheat leaves. Precis Agric 2009; 10: 34-44.
[http://dx.doi.org/10.1007/s11119-008-9082-0]

[109] Scholes JD, Rolfe SA. Chlorophyll fluorescence imaging as tool for understanding the impact of fungal diseases on plant performance: A phenomics perspective. Funct Plant Biol 2009; 36: 880-92.
[http://dx.doi.org/10.1071/FP09145]

[110] Calderón R, Navas-Cortés JA, Zarco-Tejada PJ. Early Detection and Quantification of Verticillium Wilt in Olive Using Hyperspectral and Thermal Imagery over Large Areas. Remote Sens 2015; 7: 5584-610.
[http://dx.doi.org/10.3390/rs70505584]

[111] Huang W, Lamb DW, Niu Z, Zhang Y, Liu L, Wang JJ. Identification of yellow rust in wheat using *in-situ* spectral reflectance measurements and airborne hyperspectral imaging. Precis Agric 2007; 8: 187-97.
[http://dx.doi.org/10.1007/s11119-007-9038-9]

[112] Thomas S, Kuska MT, Bohnenkamp D, *et al.* Benefits of hyperspectral imaging for plant disease detection and plant protection: a technical perspective. J Plant Dis Prot 2018; 125: 5-20.
[http://dx.doi.org/10.1007/s41348-017-0124-6]

[113] Barbedo JGA, Tibola CS, Fernandes JMC. Detecting *Fusarium* head blight in wheat kernels using hyperspectral imaging. Biosyst Eng 2015; 131: 65-76.
[http://dx.doi.org/10.1016/j.biosystemseng.2015.01.003]

[114] Arens N, Backhaus A, Döll S, Fischer S, Seiffert U, Mock HP. Non-invasive presymptomatic detection of *Cercospora beticola* infection and identification of early metabolic responses in sugar beet. Front Plant Sci 2016; 7: 1377.

[http://dx.doi.org/10.3389/fpls.2016.01377] [PMID: 27713750]

[115]   Mahlein AK, Kuska MT, Behmann J, Polder G, Walter A. Hyperspectral sensors and imaging technologies in phytopathology: State of the art. Annu Rev Phytopathol 2018; 56(1): 535-58.
[http://dx.doi.org/10.1146/annurev-phyto-080417-050100] [PMID: 30149790]

[116]   Singh A, Poshtiban S, Evoy S. Recent advances in bacteriophage based biosensors for food-borne pathogen detection. Sensors (Basel) 2013; 13(2): 1763-86.
[http://dx.doi.org/10.3390/s130201763] [PMID: 23364199]

[117]   Byrne B, Stack E, Gilmartin N, O'Kennedy R. Antibody-based sensors: Principles, problems and potential for detection of pathogens and associated toxins. Sensors (Basel) 2009; 9(6): 4407-45.
[http://dx.doi.org/10.3390/s90604407] [PMID: 22408533]

[118]   Bilkiss M, Shiddiky MJA, Ford R. Advanced diagnostic approaches for necrotrophic fungal pathogens of temperate legumes with a focus on *Botrytis* spp. Front Microbiol 2019; 10: 1889.
[http://dx.doi.org/10.3389/fmicb.2019.01889] [PMID: 31474966]

[119]   Nugaeva N, Gfeller KY, Backmann N, *et al.* An antibody-sensitized microfabricated cantilever for the growth detection of *Aspergillus niger* spores. Microsc Microanal 2007; 13(1): 13-7.
[http://dx.doi.org/10.1017/S1431927607070067] [PMID: 17234032]

[120]   Ray M, Ray A, Dash S, *et al.* Fungal disease detection in plants: Traditional assays, novel diagnostic techniques and biosensors. Biosens Bioelectron 2017; 87: 708-23.
[http://dx.doi.org/10.1016/j.bios.2016.09.032] [PMID: 27649327]

[121]   Dutse SW, Yusof NA, Ahmad H, Hussein MZ, Zainal Z, Hushiarian R. DNA-based Biosensor for Detection of Ganoderma boninense, an Oil Palm Pathogen Utilizing Newly Synthesized Ruthenium Complex (Ru(phen)$_2$(qtpy)]$^{2+}$ Based on a PEDOT-PSS/Ag Nanoparticles Modified Electrode. Int J Electrochem Sci 2013; 8: 11048-57.

[122]   Fang Y, Umasankar Y, Ramasamy RP. Plant volatile sensor: Enzymatic transducer for selective and sensitive determination of methyl salicyalte. Meeting Abstracts; The Electrochemical Society: Pennington, NJ, USA.

[123]   Jansen RMC, Wildt J, Kappers IF, Bouwmeester HJ, Hofstee JW, van Henten EJ. Detection of diseased plants by analysis of volatile organic compound emission. Annu Rev Phytopathol 2011; 49(1): 157-74.
[http://dx.doi.org/10.1146/annurev-phyto-072910-095227] [PMID: 21663436]

[124]   Cellini A, Biondi E, Buriani G, *et al.* Characterization of volatile organic compounds emitted by kiwifruit plants infected with *Pseudomonas syringae* pv. *actinidiae* and their effects on host defences. Trees (Berl) 2016; 30: 795-806.
[http://dx.doi.org/10.1007/s00468-015-1321-1]

CHAPTER 4

Machine Learning for Precision Agriculture: Methods and Applications

Ennio Ottaviani[1,2,*], Enrico Barelli[1] and Karim Ennouri[3,4]

[1] *OnAIR Ltd, Genoa, Italy*

[2] *Department of Mathematics, University of Genoa, Genoa, Italy*

[3] *Technopark of Sfax, Sfax, Tunisia*

[4] *Olive Tree Institute, Sfax, Tunisia*

Abstract: Agriculture plays a critical role in the global economy, and pressure on agricultural systems will continue to increase as the world's population grows. Modern agricultural techniques should take into account both the increased need for efficiency and the challenges posed by climate change, which together define the competing needs for sustainable farming and increased food production. Precision agriculture (PA) refers to the use of both advanced sensor technologies and state-of-the-art data analysis techniques in order to develop data-driven decision support systems. PA can help farmers to optimize crop management through accurate yield prediction and the timely detection of plant diseases and pests. Similar techniques and sensors to those used in precision agriculture can be used in the management and monitoring of livestock or fish farms, which this paper will introduce for completeness. A survey of machine learning methods will be presented in order to provide researchers and end-users with an up-to-date starting point for their projects and use-cases.

Keywords: Artificial Intelligence, Crop Management, Data Analysis, Livestock Management, Machine Learning, Precision Agriculture, Smart Farms, Soil Management, Statistical Prediction.

INTRODUCTION

Precision agriculture (PA) is a broad term encompassing methods and enabling technologies that can provide scientific, sound and reliable decision support systems for farmers at a great level of granularity and detail. The need for such systems arises from the world's increasing population and the new challenges that climate change poses to crops. Moreover, PA technologies can improve farm management by providing timely, detailed and site-specific information on farm

---

* **Corresponding author Ennio Ottaviani:** OnAIR Ltd, Genoa, Italy, Department of Mathematics, University of Genoa, Genoa, Italy; E-mail:ennio.ottaviani@onairweb.com

Karim Ennouri (Ed.)

production. As a result, the costs of running a farming enterprise can be reduced and profits can be increased. However, the adoption of this new kind of approach has been sluggish and far from universal, as reported by Schimmelpfennig *et al.* in [1]. Indeed, farmers frequently implement PA solutions sequentially, starting from a few simple modules, even when the adoption of a whole PA package would be more advantageous. This could, in turn, hinder the adoption of the most innovative solutions, in that the early savings may be deemed sufficient and further risk may be unwanted. A study by Najafabadi *et al.* [2], which explored the main challenges involved in implementing PA technologies through the analysis of questionnaires, found that utilizing such systems could increase profitability and production and reduce the environmental impact of chemicals. One key concern is that the implementation of such technologies requires a substantial initial investment, both financial and in terms of the time spent learning how to use them; moreover, further time will pass before this investment yields an optimal return, an aspect that is of particular concern to small farmers. While that study mainly explored the Iranian situation, similar concerns were discussed in a study by Kritikos *et al.* [3] conducted in the European Union. Further challenges concern data quality and educational, demographic and technical problems.

The present work looks at several studies concerning various areas of application of PA technologies, the aim being to identify cases of the successful use of these methodologies in order to facilitate both further implementation and research. Other surveys can be found in the literature (such as Alreshidi *et al.* [4] and Liakos *et al.* [5]). The first of these concerns itself with tools of the Internet of Things (IoT); it provides extensive examples of possible sensors and infrastructure, and concludes by suggesting the best practices for implementing a complete pipeline for sustainable precision farming. The second survey [5] provides an extensive list and very brief descriptions of previous studies, with the aim of highlighting the most frequently utilized learning methods and the most commonly explored fields of application. One of its findings is that 71% of publications on PA are related to crops and soil applications, while the rest of them focus on aquaculture and livestock management. Fig. (**1**) shows data on the subfields of crop and soil-related domains reported in the survey. The present work will explore a limited subset of possible applications of PA to crops and the most successful methods used, as found in the literature. For completeness, a brief introduction to aquaculture and livestock management is provided.

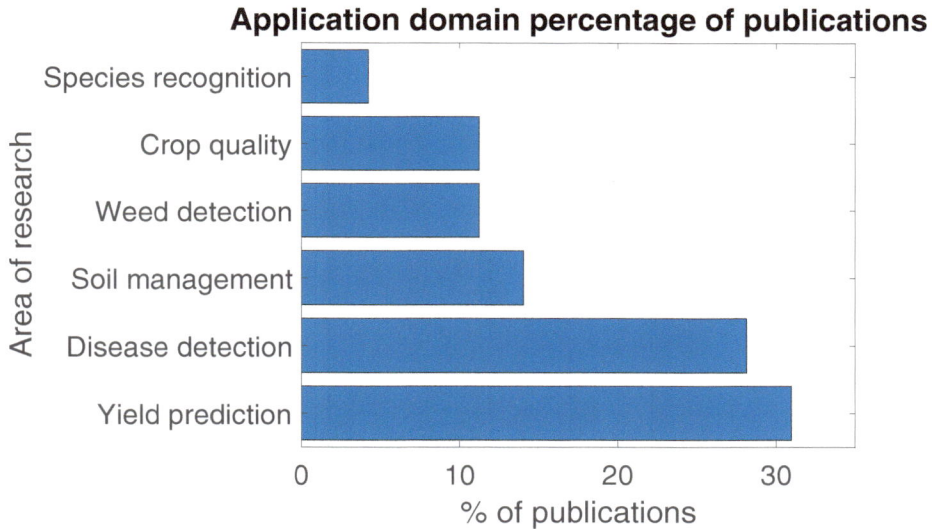

Fig. (1). Application domains by the percentage of publications.

MACHINE LEARNING AND COMPUTER VISION OVERVIEW

In order to better understand the PA applications presented and the cases of their use, the following sections will briefly introduce the terminology of machine learning and computer vision and provide terminology and pointers to the literature.

Statistics and Machine Learning

Statistical learning methods involve finding a suitable function (*i.e.*, a statistical model) to explain the behavior of a dataset. The model is built by optimizing a suitable performance measure and then quantifying how well the model describes all the available data (the training set). The data consist of a set of examples, in which an individual example is described by a set of features. A feature may be categorical, binary, ordinal or numeric. The final aim of the learning process is to generalize the descriptive capability of the model to new data, while avoiding overfitting. Statistical learning methods have led to many successful applications in various fields, such as robotics, medicine, biology and many others. While their application in agriculture is more recent, the effectiveness of the approach is beyond doubt. Finally, the term "statistical learning" is often confused with the more common term "machine learning". Indeed, there is no clear difference between them. Machine learning is a sub-field of artificial intelligence that studies algorithms that allow a computer to learn to perform a given task. Here, we will use the two terms as if they were interchangeable. The typical flow of a statistical/machine learning approach is depicted in Fig. (**2**).

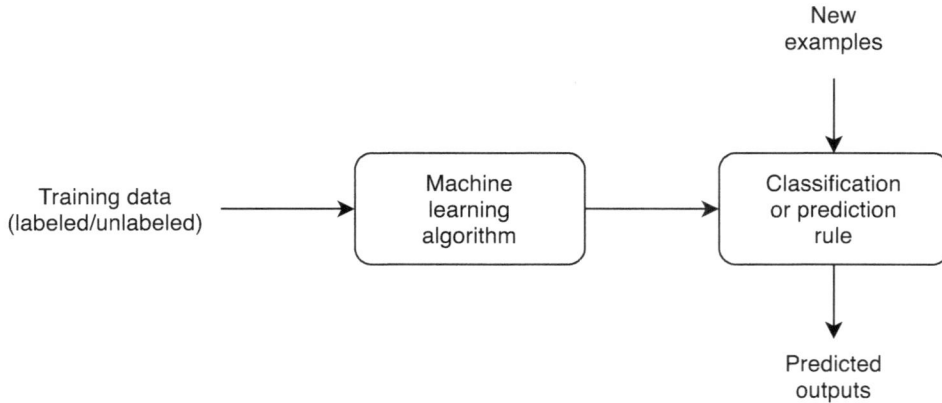

Fig. (2). Typical flow of a machine learning approach.

Machine learning methods are typically classified in different broad categories, depending on the type of learning (supervised or unsupervised), the learning model (classification, regression, clustering, dimensionality reduction, ...), and the specific learning algorithm employed to realize the task. Many textbooks, such as [6] by Hastie *et al.*, present good reviews of all these methods, which include generalized linear and logistic regression, Support Vector Machines (SVM), tree-based methods (Bagging, Random Forests and Adaptive Boosting) and neural networks (Multi-Layer Perceptrons [MLPs]).

Computer Vision and Deep Learning

Machine learning methods need a feature representation in order to describe the process to be analyzed. Features typically come from sensors or other observations. When input data consist of images, computer vision is needed in order to produce a set of features for each image location. Image features can capture how the colors (or gray levels) of the image are distributed around a given point and enable a machine learning algorithm to be constructed in order to analyze local image patterns. Given a task to be solved, computer vision researchers look for the most suitable feature descriptor, using their experience as guidance. Many textbooks present a comprehensive review of the different feature descriptors proposed to capture local image properties, as [7] by Davies. The close connection between computer vision and machine learning in the case of a classic object detection task (car recognition) is shown in Fig. (**3**).

Training
1. Obtain training data
2. Define features
3. Define classifier

Testing
1. Slide window
2. Score by classifier

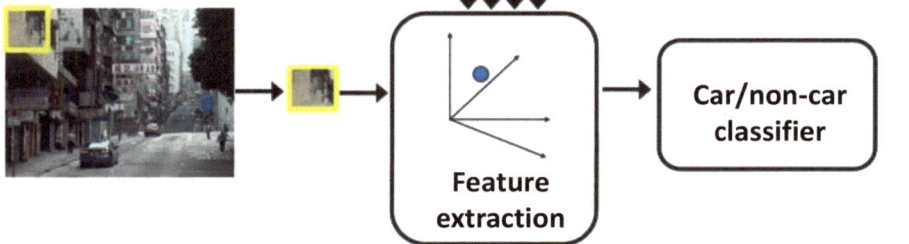

Fig. (3). Computer vision and statistical learning for car recognition.

When dealing with images, it is often very hard to find a suitable feature descriptor for a given problem; therefore, a method for learning the best features to use together with the statistical model can be very useful. Deep Learning (DL) achieves this goal by using the classic multi-layer perceptron model by increasing and modifying the number of layers. The result is a single network that has an input layer connected directly to image pixels, a possibly high number of hidden layers with particular structures, and an output layer providing the final result. DL models are perfectly suited to dealing with images, as they do not need any assumption regarding the features, and they learn directly from the pixel values.

Of the various DL models, about which a good introductory textbook is [8] by Goodfellow *et al.*, perhaps the most popular is the Convolutional Neural Network (CNN). In this network, the first hidden layers are substituted by image convolutions with a set of digital filters, the coefficients of which are learned during the training process. Convolutional filters are well-known in computer vision as a way to get image features, as shown in the book of Davies [7].

It must be stressed that, in the CNN model, the convolutional filters are not defined by an experienced practitioner; rather, they are learned from the training data. After constructing the convolutional layers, a standard MLP network performs image classification. CNNs gained wide popularity by solving recognition problems (*e.g.*, handwritten character recognition) with outstanding accuracy, as reported by LeCun *et al.* [9]. A typical CNN architecture for handwritten character recognition (LeNet-5) is shown in Fig. (**4**).

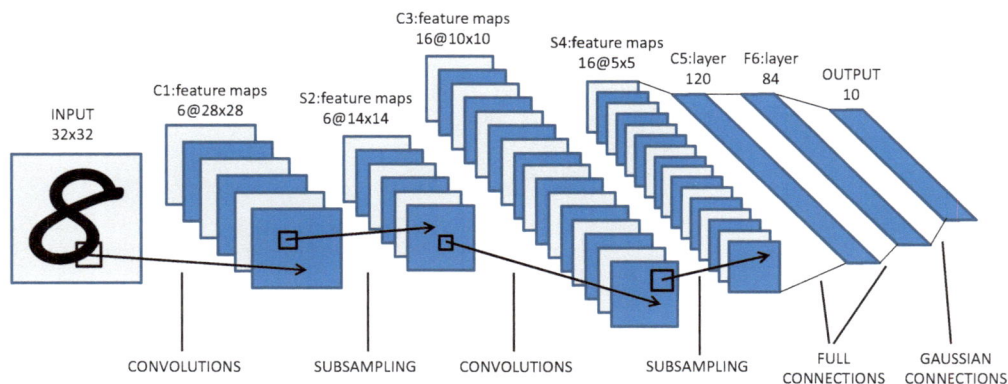

Fig. (4). CNN architecture (LeNet-5) for optical character recognition (adapted from [9]).

Today, CNNs for complex image recognition tasks handle larger images (even in color or of multi-spectral type) and use a larger number of layers. Computational problems can be partially solved by using massive parallel computing boards (GPUs), which are often present in modern computers, or dedicated hardware units. Many CNNs are already available for generic image recognition tasks (*e.g.*, AlexNet, ResNet, GoogLeNet). These can be applied to a new problem by modifying and training only the last MLP layers, in order to match the final output with the desired one (transfer learning). This greatly reduces the need for training from scratch in order to use a new deep learning model, thereby allowing fast prototyping of new applications in different fields.

APPLICATIONS OF PRECISION AGRICULTURE AND USE-CASES

In the following sections, PA applications and cases of their use will be presented. Various domains within PA will be explored along with different machine learning models and computer vision approaches, ranging from the very simple to the most complex.

Crop Yield Prediction and Evaluation

Yield prediction is one of the most critical domains in PA, as it allows crop supply to be matched with demand and crops to be correctly managed in order to increase productivity through sensible planning. In Panayi *et al.* [10], the authors used a functional regression framework to predict the yield of *Agaricus Bisporus* (button mushroom), and to quantify the contribution of the environmental variables that could be precisely controlled in an indoor mushroom farm. The optimal calibration of habitat throughout the growth cycle of the mushrooms proved able to increase the farm yield. These authors examined time series of uneven length, which comprised all sensor readings recorded during past growth cycles, and used these jointly in their subsequent analysis. The length of crop cycles may vary greatly, as each cycle partly depends on uncontrollable external factors. This variability, however, can cause serious problems for the data analyst. The authors' solution was based on a functional representation of data based on B-splines, which are an advanced curve-fitting technique; for an introduction to spline approximation techniques, the reader should refer to the textbook by Hastie *et al*. [6]. The main result was that it was indeed possible to optimize mushroom crop yields by carefully implementing a predictive method which, by utilizing the data acquired in order to solve particular problems, enabled the environmental variables to be carefully calibrated.

In the work by You *et al*. [11], a method of predicting outdoor crop yields in the absence of high-quality, well-organized data is proposed. These authors state that methods based on on-site sensor data and equipment provide optimal results, but are expensive and readily available only in the most developed countries. However, it is in developing countries that methods that produce accurate results on the basis of publicly available data are most needed to improve food security. The method described uses remote sensing data (satellite images) that are globally available and relatively inexpensive. The authors propose a deep learning pipeline that uses a sequence of pre-harvest images of a region of interest in order to predict an average yield per unit area in a given geographical region. In their study, the authors propose the in-season prediction of soybean yield throughout the USA as an example. Their forecasts outperformed those of the US Department of Agriculture in the year of publication, demonstrating that very accurate yield prediction can be achieved by using relatively cheap remote sensing data. Remote sensing technologies can also be applied to the assessment of crop condition, which, in combination with yield estimation, is an important factor to monitor during the growth cycle in order to avoid diminishing the production levels. A review of methods and measures in this domain is given in the work by Ennouri *et al*. [12].

In several industrial applications, decision-making is aided by simulation software, which, in conjunction with statistical analysis, can help to reach optimal decisions in different scenarios. Memic *et al*. present in [13] an open-source extension to GeoSim, a plug-in to the QGIS [14] software, which is also open-source, is developed, the goal being to determine the optimal quantity of nitrogen to administer to each individual lot of maize crops partitioned in a regular grid. Applying this prescription can both increase profits and reduce environmental water pollution by limiting the use of unneeded fertilizer.

The model was tested in two fields located in the USA and Germany. The program generates yield and nitrogen levels in the soil at harvest-time in user-specified grids and weather conditions. The economic optimizer component of the software allows the user to enter the selling price of maize, the cost of nitrogen and other economic factors. It then computes the Marginal Net Return (MNR) over a range of nitrogen rates in user-specified weather conditions. The seasonal MNRs are then used to compute the nitrogen prescription that maximizes the long-term MNR over the user-selected seasons of weather data, which may be future forecasts, thereby facilitating the planning of future operations. The simulation also allows the user to obtain estimates of which environmental factors most strongly influence the outcome.

The results obtained by retrofitting historical data showed that it was possible to greatly optimize crop yield with respect to a uniform use of fertilizer, up to 48% reduction in nitrogen rate and 11% increase in MNR in the German use-case, and respectively 9% and 6% in the US use-case. The main conclusion was that crop simulation models could indeed provide a helpful decision support system and that they could be further improved with more accurate and dense input data directly sampled on-site.

In this section, several methods of predicting and assessing crop yield on the basis of various fields of statistics have been presented. If different data sources and different mathematical and information technology tools are used, the methods cited should provide readers with starting points for their own research or uses. In the following sections, various different domains of application of PA will be explored.

Soil Management and Description

In order to apply precision agriculture practices more accurately, detailed and granular knowledge of the soil composition is needed. The main goal is to be able to estimate the within-field variation in nutrients and soil characteristics, in order to micro-manage interventions calibrated on specific subsections of the target field. In order to do so, farmers need sufficient information. While some data are provided at very high resolution by sensors directly attached to farm equipment (*e.g.*, tractors), other kinds of data are much sparser; for example, soil information may require expensive sampling and chemical analysis in a laboratory. This sparse information, therefore, needs to be extrapolated as accurately as possible to the whole field. In geostatistics, a specific sub-field of statistics, this kind of analysis has been very thoroughly studied. Commonly called kriging, it concerns the study of optimal estimators of spatially dependent data.

A brief description of this kind of technique is given in the work of Oliver *et al.* [15], which also provides an example case study in which thematic maps obtained by means of kriging are produced. These maps highlight the sand content of the soil, which is an important piece of information in precisely planning the application of farmyard manure and seed use rate. In the example provided, the subsequent improvements in yield and reductions in cost prompted farmers to vary their seeding rates automatically on the basis of regular updates of the maps and to study the use of satellite images in order to apply precision agriculture even further. For an in-depth theoretical analysis of kriging, the reader should refer to a textbook such as [16] by Chiles and Delfiner.

In a paper by Coopersmith *et al.* [17], the authors demonstrate how a fairly basic Machine Learning algorithm can be used, without needing sensors placed directly in the field, in order to correctly classify the state of soil with respect to its readiness to be worked. The study described in the paper used only data that were publicly available in the US state of Illinois to develop a model to predict whether soil dryness, as defined by users with respect to their specific applications, was at a sufficient level to allow agricultural work to be carried out. The study involved a farm site located in Urbana, Illinois, and the hypothesis was that knowing the state of dryness of the soil in advance could help farmers plan their activities and avoid damage to their equipment, as may happen when working in unfavorable soil conditions. The methods used were not particularly recent, and the dataset was not rich. Nevertheless, the study showed that, through a careful choice of variables, free data and open-source software, it was possible to build a model that, with just one month of data in the training phase, was able to produce satisfactory results in comparison with those yielded by the manual sampling of the soil and assessment by a human expert. This seems encouraging, in that it shows that useful systems to support decision-making can be constructed even when significant financial investments are not available.

Disease, Weed, Pest Detection and Management

In a paper by Meiser *et al.* [18], the authors tried to develop an optimal pest-management policy for cotton fields in California's San Joaquin Valley region. The goal was to devise a data-driven strategy for pesticide use that would maximize profits by reducing the costs of purchasing chemicals while improving crop yields. Although reducing pesticide use also has a beneficial effect on the environment, this aspect was not explicitly taken into account in the authors' decision-making function, and the farmers were not fully convinced to adopt the optimal policy that would maximize their short-term profits. Data were collected from 1498 commercial cotton crops and referred to more than 10 years of crop yields, pesticide application and pest density. The method developed used Markov

Decision Processes. This approach defines both possible actions and possible states and the transition between states, in order to devise optimal actions to facilitate the transition between those states that are best with regard to a given cost function. In the case described, the function calculated profits on the basis of cotton prices, labor costs and pesticide costs. The transition probabilities over a given time were estimated by means of Bayesian ordered logistic regression, and the difference in yield and costs between different fields and cotton species through Bayesian hierarchical models. The results defined an optimal strategy. However, on carrying out a retrospective analysis, the authors noted that the farmers tended to rely only on their experience, and rarely pursued the data-driven optimum, thus highlighting the importance of precision agriculture practices.

In a work by Smaoui *et al.* [19], statistical and artificial intelligence methods were compared in a more biologically-oriented study than those described above. The goal was to devise an optimization strategy to increase the anti-pathogenic capabilities of a new bacterial strain, the use of which would enable microbial pathogens to be controlled biologically without the use of expensive and environmentally harmful chemicals. Through carefully designed experiments and biological and statistical methods, the authors managed to determine the optimal mean composition for the growth of *Streptomyces sp. TN71* in order to enhance its anti-*A. tumefaciens* capability. *A. tumefaciens* is a bacterium that causes various neoplastic diseases of plants, making several food commodities unfit for human consumption. The study show that statistical learning methods could increase agricultural efficacy at the biological level, even before higher-level decisions regarding crop management are taken.

Useful information on plant diseases can be automatically extracted through the image analysis of leaves. Particular uses have been found for multispectral or hyperspectral images, each pixel of which contains information from wavelengths that are outside the visible light spectrum. The difference between multispectral and hyperspectral images lies in their sampling methods; multispectral images usually collect information from discrete, narrow bands, usually from the visible range up to the near-infrared or infrared, while hyperspectral images offer much higher spectral resolution and yield information at much more fractionated intervals. In the paper by Rumpf *et al.* [20], hyperspectral images and Support Vector Machines were used to analyze sugar beet leaves inoculated with various pathogens, in order to create an automatic method of detecting diseases at an early stage. Support Vector Machines are powerful tools, and constituted state of the art for more than a decade in various domains of application, before the recent advent of deep learning. The textbook [6] devotes an entire chapter to Support Vector Machines, to which the reader should refer.

In the above-mentioned study, data were collected from healthy sugar beet plants and plants inoculated with three different pathogens: *Cercospora beticola*, *Uromyces betae* or *Erysiphe betae*; these data were recorded for 21 days after inoculation. Hyperspectral images were recorded and disease severity was evaluated and classified. Several vegetation indexes were calculated from the raw spectra at each sampling time. Even though the reported performances varied from disease to disease and over time, the study demonstrated that it was possible to construct an automatic system to detect sugar beet diseases even in the pre-symptomatic stage. In the field of hyperspectral imaging, it is still difficult to obtain the huge datasets that are used in other image classification tasks. Indeed, the available data are mostly provided by satellites and are limited in terms of both resolution and amount, which makes it hard to train deep learning systems. However, when bigger and higher quality datasets become available, even better performances may be achieved by means of the current state-of-the-art methods, such as Convolutional Neural Networks (CNNs). CNNs are being very successfully applied to classical RGB images. For example, in the work by Sladojevic *et al.* [21], the authors developed a CNN-based system for recognizing 13 different types of plant diseases by comparing healthy leaves with leaves from diseased plants. The authors describe in detail the process of dataset acquisition and enrichment and the architecture of the network used. The paper provides non-experts with a good introduction to the application of deep learning techniques. Moreover, in their future work, the authors propose to develop a mobile application that would allow farmers to photograph plants on their smartphones and to receive a diagnosis of the plant disease in real-time, which would support decision-making. This kind of technology would help even untrained individuals hired for other kinds of work, such as maintenance, or perhaps even surveillance cameras, to continuously monitor fields for possible diseases.

Phenology Recognition

The ability to predict phenomena related to crop phenology, *i.e.* classify which kind of crops an image contains or at which stage of growth a single crop is, is useful for a variety of monitoring and decision-making applications of government and private sectors, such as crop insurance, land rental and supply-chain logistics, as shown by Cai *et al.* [22]. In this paper, the authors propose a method of fusing multispectral data from several satellites, such as multiple Landsat missions, and data from the US Department of Agriculture field parcels. This data-fusion approach enabled the authors to build a dataset spanning 15 years, without missing data. This approach is interesting because it enables a more robust study of problems of in-season forecasting. The use case presented was field-level crop classification, which the authors performed with a high level of in-season accuracy.

In another example by Bauer *et al.* [23], aerial imaging from fixed-wing light aircraft equipped with sensors able to capture Normalized Difference Vegetation Index (NDVI) was used to build an analytic platform for the in-season phenotyping of lettuce fields. As NDVI correlates well with leaf-area index and biomass, it was chosen by the authors for yield-related field phenotyping. In partnership with a large grower, the authors managed to acquire a large annotated dataset in which each lettuce head was identified by a 20x20 pixel bounding box; more than 100,000 lettuce heads were annotated, which gives an idea of a large amount of work required in order to obtain a dataset from which such a system can be built. From this dataset, the authors created a pipeline comprising: pre-processing to stabilize the image signal, a CNN for lettuce segmentation, a clustering algorithm to define and then classify lettuce size, which correlates with the growth stage, and a visualization platform to inspect the field- and yield-related features. All the resulting information was GPS-tagged in order to provide the user with precise, geo-referenced information on the fields monitored, thereby allowing a sound harvesting strategy to be elaborated on the basis of the growth stage of the lettuce. Once such a software platform has been established, the cost of the infrastructure lies in maintaining the aircraft and periodically acquiring images along planned flight routes. The authors state that it should not be too difficult to extend their work to other kinds of crops, such as rice or wheat, and thus to some of the largest agricultural enterprises. The platform is open-sourced and, at the time of writing, can be found at https://github.com/Crop-Phenomic--Group/Airsurf-Lettuce/releases. The reader should use this link to see an example of how such a system can be created, and hence to quantify the effort needed to emulate such results.

An interesting application of deep learning to the classification of phenological stages of several types of plants on the basis of visual data is presented in Yalcin *et al.* [24]. Images are captured every half an hour by cameras mounted on the ground at agro-stations in Turkey, as part of an agriculture monitoring network system. A pre-trained Convolutional Neural Network architecture (AlexNet) is employed to automatically extract the features of images and to classify different phenological stages of the growth of wheat, barley, lentils, cotton, peppers and corn. The performance of the proposed approach has been evaluated by comparing the results obtained through the AlexNet with those obtained by means of handcrafted feature descriptors, and has shown significant improvements.

Finally, it should be noted that most existing approaches to crop classification concentrate on single-time measurements. However, as crops change their reflective characteristics over time, classification performance depends on the particular time of observation. However, as these characteristics change in a systematic manner, a multi-temporal approach can be utilized to overcome this problem. In one such case by Russwurm *et al.* [25], a time-dependent deep learning model (Long and Short Term Memory, LSTM [8]), was used to extract temporal characteristics from a sequence of SENTINEL 2A satellite images. A large study area in Germany with ground truth annotations provided by public authorities was used for training and evaluation of 20 crop classes. The results seem promising, in comparison with non-temporal approaches.

Livestock and Aquaculture Applications

Techniques that are fairly similar to those described for agriculture can be used in livestock farming or aquaculture, two areas which also provide huge quantities of food. In these areas, it is essential to improve both food safety and production, while also ensuring sustainability and animal welfare and reducing the environmental impact of these activities. A survey on precision livestock farming by Banhazi *et al.* [26] has documented some specific areas in which such techniques can be utilized. These include: improving and documenting animal welfare, reducing greenhouse gas emissions, improving the environmental performance of farms, and facilitating the economic stability of rural areas.

As noted in a contribution by Morgan-Davies *et al.* [27], individual electronic identification (EID) of animals has become widespread, mostly in the form of ear-tags. In Europe, this has been mandatory since 2004. The authors point out that, while this practice is often perceived as an additional burden on the farmer, it is a prerequisite to precision livestock farming, as it allows management at the unit level. In their study, the authors monitored two sheep flocks for 3 years; one was managed conventionally and the other by means of a precision farming protocol. In the latter flock, weighing was performed automatically by means of a weighing crate connected to the EID; the data collected were used to divide the flock into groups on the basis of phenotypic attributes (*e.g.*, weight change). In the traditional setting, by contrast, the flock was divided according to the shepherd's subjective assessment. The goal was to decide how to optimally feed different groups during winter, how to administer worming treatment, and to assess the amount of work saved (if any) through precision farming. The main results showed that it was actually possible to achieve significant savings and, in particular, to reduce worming treatments and to apply them more selectively. This study reveals that regulators can implement facilitating mechanisms to encourage precision farming, but that often such measures are not adequately supported by incentives or proper education, which are necessary for their complete implementation.

A review of commercial software and its respective fields of application is provided by Banhazi *et al.* [28]; some examples will be reported in order to give the reader an idea of the possible solutions. One type of software reviewed offers a system of continuous approximate weighing obtained by correlating images from an indoor pig farm with manual weighing; such a system can monitor pigs at an individual level. A feeding system that can monitor how much dry food is served to each individual is proposed, together with various sensors to monitor environmental variables inside a farm. These technologies differ from those used in agriculture, which are described in greater detail in the previous sections, only in the details of their implementation.

Aquaculture is another field in which the concept of precision can be applied. However, it poses unique challenges, such as difficulties in deploying a wireless communication network under water, the need for waterproof insulation for sensors, bio-fouling phenomena and maintenance difficulties, as pointed out in Parra *et al.* [29]. In that paper, the authors propose a review of the state of the art of the sensors used in precision aquaculture. The reader should refer to the original paper for details of the requirements and design factors to respect during the instrumentation of an aquaculture facility. Example applications and case studies aimed at solving specific problems, such as biomass monitoring, control of feed delivery and parasite monitoring, are given in Fore *et al.* [30]. An interesting example of how computer vision can be combined with statistical learning to monitor fish welfare in salmon farms, as indicated by jumping and splashing, is given in Jovanovic *et al.* [31]. In their study, the authors used an unmanned aerial vehicle (UAV) to obtain videos of salmon farms. They then developed an algorithm that combined traditional computer vision feature descriptors, such as Fast Retina Keypoint Descriptors (FREAK [32],) or Vectors of Locally Aggregated Descriptors (VLAD [33],) with a Support Vector Machine (SVM) in order to detect and classify splashes in the fish farm. The number of times that fish jump above the water, causing splashes at the surface, is closely related to fish welfare and stress levels. This video processing technique was used on an embedded system placed at the fish farm, along with a fixed-position camera, and provided suitable outdoor casing. With appropriate data transmission, a real-time surface activity sensor could be created.

CONCLUSION

The present paper explores various domains of the application of machine learning methods to precision agriculture. In each domain, several kinds of algorithms have been proposed, from very classical ones, such as kriging, to the most recent ones, such as Convolutional Neural Networks. Examples of the hardware and sensors needed are provided. With the help of the papers cited as examples, readers should be able to identify the class of problems closest to their interests and evaluate the kind of method most applicable to the situation, according to their mathematical and statistical background. Pointers to the literature on the different algorithms have been given in order to provide the researcher or practitioner with a good starting point for further work.

By applying machine learning methods to sensor and image data, farm management systems can gain useful information to support their decisions and actions, the final aim being to improve farm management. In the near future, it is expected that the use of these methods will become more widespread. As yet, however, many individual solutions are not adequately connected with the decision-making process, as indeed was the case in other domains, such as factory automation. This integration of data recording, data analysis, statistics, machine learning and final decision-making would allow precision agriculture to further increase both production levels and product quality.

CONSENT FOR PUBLICATION

Not applicable.

CONFLICT OF INTEREST

The author(s) confirms that there is no conflict of interest.

ACKNOWLEDGEMENTS

Declared none.

REFERENCES

[1]   Schimmelpfennig D, Ebel R. Sequential adoption and cost savings from precision agriculture. J Agric Resour Econ 2016; 41(1): 97-115.

[2]   Najafabadi MO, *et al.* A bayesian confirmatory factor analysis of precision agricultural challenges. African Journal of Agricultural Research 2011; 6(5): 1219-25.

[3]   Kritikos M. Precision agriculture in Europe, legal social and ethical considerations European Parliamentary Research Service, Scientific Foresight Unit 2017.

[4]   Alreshidi E. Smart sustainable agriculture solution underpinned by IoT and artificial intelligence. International Journal of Advance Computer Science and Application 2019; 10(5): 93-102.

[5]   Liakos KG, Busato P, Moshou D, Pearson S, Bochtis D. Machine learning in agriculture: A review. Sensors (Basel) 2018; 18(8): 2674.
[http://dx.doi.org/10.3390/s18082674] [PMID: 30110960]

[6]   Hastie T, Tibshirani R, Friedman J. The elements of statistical learning. Springer Series in Statistics 2001.
[http://dx.doi.org/10.1007/978-0-387-21606-5]

[7]   Davies ER. Computer Vision Principles, algorithms, applications, learning. Academic Press 2017.

[8]   Goodfellow I, Bengio J, Courville A. Deep learning. The MIT Press 2016.

[9]   LeCun Y, Botton L, Bengio Y, Haffner P. Gradient-based learning applied to document recognition. Proc IEEE 1998; 86(11): 2278-324.
[http://dx.doi.org/10.1109/5.726791]

[10]   Panayi E, Peters GW, Kyriakides G. Statistical modelling for precision agriculture: A case study in optimal environmental schedules for *Agaricus Bisporus* production *Via* variable domain functional regression. PLoS One 2017; 12(9)e0181921
[http://dx.doi.org/10.1371/journal.pone.0181921] [PMID: 28961254]

[11]   You J, Li X, Low M, Lobell D, Ermon S. Deep gaussian process for crop yield prediction based on remote sensing data. Proceedings 31st AAAI Conference on Artificial Intelligence.

[12]   Ennouri K, Kallel A. Remote Sensing: an advanced technique for crop condition assessment. Math Probl Eng 2019; 1: 1-8.
[http://dx.doi.org/10.1155/2019/9404565]

[13]   Memic E, Graeff S, Claupein W, Batchelor WD. GIS-based spatial nitrogen management model for maize: short and long-term marginal net return maximising nitrogen application rates. Precis Agric 2019; 20: 295-312.
[http://dx.doi.org/10.1007/s11119-018-9603-4]

[14]   QGIS Development Team. QGIS Geographic Information System Open Source Geospatial Foundation Project 2019 2019.

[15]   Oliver MA. Precision agriculture and Geostatistics: how to manage agriculture more exactlySignificance, the Royal Statistical Society 2013.
[http://dx.doi.org/10.1111/j.1740-9713.2013.00646.x]

[16]   Chiles JP, Delfiner P. Geostatistics: modeling spatial uncertainty. Wiley Series in Probability and statistics 1999.

[17]   Coopersmith E, Minsker BS, Wenzel CE, Gilmore BJ. Machine learning assessments of soil drying for agricultural planning. Comput Electron Agric 2014; 104: 93-104.
[http://dx.doi.org/10.1016/j.compag.2014.04.004]

[18]   Meisner MH, Rosenheim JA, Tagkopoulos I. A data-driven, machine learning framework for optimal pest management in cotton. Ecosphere 2016; 7(3)e01263
[http://dx.doi.org/10.1002/ecs2.1263]

[19]   Smaoui S, Ennouri K, Chakchouk-Mtibaa A, *et al.* Modeling-based optimization approaches for the development of Anti-Agrobacterium tumefaciens activity using Streptomyces sp TN71. Microb Pathog 2018; 119: 19-27.
[http://dx.doi.org/10.1016/j.micpath.2018.04.006] [PMID: 29626659]

[20]   Rumpf T, *et al.* Early detection and classification of plant diseases with Support Vector Machine based on hyperspectral reflectance. Comput Electron Agric 2010; 74(1): 91-9.
[http://dx.doi.org/10.1016/j.compag.2010.06.009]

[21]   Sladojevic S, Arsenovic M, Anderla A, Culibrk D, Stefanovic D. Deep neural networks based recognition of plant diseases by leaf image classification. Comput Intell Neurosci 2016; 20163289801
[http://dx.doi.org/10.1155/2016/3289801] [PMID: 27418923]

[22]   Cai Y, *et al.* A high-performance and in-season classification system of field-level crop types using time-series Landsat data and a machine learning approach. Remote Sens Environ 2018; 210: 35-47.
[http://dx.doi.org/10.1016/j.rse.2018.02.045]

[23]   Bauer A, Bostrom AG, Ball J, *et al.* Combining computer vision and deep learning to enable ultra-scale aerial phenotyping and precision agriculture: A case study of lettuce production. Hortic Res 2019; 6(70): 70.
[http://dx.doi.org/10.1038/s41438-019-0151-5] [PMID: 31231528]

[24]   Yalcin H. Plant phenology recognition using deep learning: Deep-Pheno. 6th International Conference

on Agro-Geoinformatics. 1-5.
[http://dx.doi.org/10.1109/Agro-Geoinformatics.2017.8046996]

[25]   Russwurm M, Korner M. Temporal vegetation modelling using long short-term memory networks for crop identification from medium-resolution multi-spectral satellite images. IEEE Conference on Computer Vision and Pattern Recognition Workshop. 1496-504.

[26]   Banhazi TM, *et al.* Precision Livestock Farming: An international review of scientific and commercial aspects. Int J Agric Biol Eng 2012; 5(3): 1.

[27]   Morgan-Davies C, *et al.* Impacts of using a precision livestock system targeted approach in mountain sheep flocks. Livest Sci 2018; 208: 67-76.
[http://dx.doi.org/10.1016/j.livsci.2017.12.002]

[28]   Banhazi TM, Babinszky L, Halas V, Tscharke M. Precision Livestock Farming: Precision feeding technologies and sustainable livestock production. Int J Agric Biol Eng 2012; 5(4): 54-61.

[29]   Parra L, Lloret G, Lloret J, Rodilla M. Physical sensors for precision aquaculture: a review. IEEE Sens J 2018; 18(10): 3915-23.
[http://dx.doi.org/10.1109/JSEN.2018.2817158]

[30]   Fore M, *et al.* Precision fish farming: A new framework to improve production in aquaculture. Biosyst Eng 2018; 173: 176-93.
[http://dx.doi.org/10.1016/j.biosystemseng.2017.10.014]

[31]   Jovanovic V, *et al.* Splash detection in surveillance videos of offshore fish production plants. International Conference on Systems, Signals and Image Processing.
[http://dx.doi.org/10.1109/IWSSIP.2016.7502706]

[32]   Ortiz R. FREAK: Fast retina keypoint. IEEE Conference on Computer Vision and Pattern Recognition. 510-7.

[33]   Husain S, Bober M. Robust and scalable aggregation of local features for ultra large-scale retrieval. IEEE International Conference on Image Processing. 2799-803.
[http://dx.doi.org/10.1109/ICIP.2014.7025566]

CHAPTER 5

Use of Remote Sensing Technology and Geographic Information System for Agriculture and Environmental Observation

Karim Ennouri[1,*], Ennio Ottaviani[2,3], Slim Smaoui[4] and **Mohamed Ali Triki[1]**

[1] *Olive Tree Institute, Sfax, Tunisia*

[2] *OnAIR Ltd, Genoa, Italy*

[3] *Department of Mathematics, University of Genoa, Genoa, Italy*

[4] *Centre of Biotechnology of Sfax. Sfax, Tunisia*

Abstract: Information on the environment, coverage, spatial division along by way of natural resources is a condition to accomplish the objectives of biological agriculture and sustainable development. Geographic Information System (GIS) suggests a perfect setting for incorporating spatial and characteristic information on the environment and natural resources. Furthermore, remote sensing permits making accurate records on an assortment of landscape factors which are employed for both creating baseline in addition to derivative records on natural resources used for a range of agricultural actions. Innovations in predicting and telecommunication help in the valuable functioning of most advantageous land use strategies/exploitation strategies. This chapter provides a general idea about the role of remote sensing, Geographic Information System, and digital photogrammetry. Furthermore, the chapter also identifies the progress and developments in captor technology, records processing and explanation/investigation and combination of geospatial records and data.

Keywords: Agriculture, Captor, Geographic Information System, Remote Sensing.

INTRODUCTION

Nowadays, it is clear that the surface of the globe is quickly modifying due to a range of causes at limited and local degrees, with considerable consequences for persons and also for the environment. In the aim to acquire greater comprehension, studying and forecasting the transformations, remote sensing satellite pictures and images represent a vast foundation of helpful information for cover analysis and investigation.

---

[*] **Corresponding author Karim Ennouri:** Olive Tree Institute, Sfax, Tunisia, E-mail:karim.ennouri@gmail.com

Explanation and examination of remote sensing imagery engage the recognition and/or measurement of diverse targets in a picture in order to extract practical information about details [1]. There are many applications of remote sensing in the agricultural sector. In fact, Remote sensing is employed to predict the expected crop productivity and yield over a given region and decide how much of crop will be collected under particular conditions. Moreover, Remote sensing technology has been influential in the examination of different crop planting structures. This skill has been applied principally in horticulture manufacturing, where flower development patterns can be analyzed and a forecast made out of the analysis [2]. Moreover, it also plays a significant function in the assessment of plant health crop state and the extent to which the crop has withstood stress. This data is then used to determine the crop quality. Furthermore, it plays a significant function in pest detection in farmland [3].

Remotely sensed satellite examinations from space have basically revolutionized the approach in which researchers learn the atmosphere, mountain, soil, flora, sea, and other ecological aspects of the globe ground. More than fifty years of the satellite surveillances of the globe have offered realistic images and they are the foundation for a novel scientific concept: the earth-system discipline. Remote sensing methods are essential in obtaining valuable information of the globe *via* detectors. The remotely assembled records will be examined in order to acquire data about objects, regions being studied [4]. In addition, it comprises the test and explanation of the obtained records and descriptions, which are the most features that offer pertinent information for ecological researchers in checking globe reserves [5]. Multi-spectral imagery can be employed for the quantification and monitoring of resources through a specified duration of time. Remote sensing methods assist developing regions in studying possible deforestation and transformations in vegetation cover. Besides, geographic information systems are great tools when employed in geographical sciences and land exploitation domains. The energy source, as electromagnetic energy, is the critical mean necessary to pass on information from studied objective to detector. Remote sensing knowledge formulates the utilization of an extensive assortment of Electro-Magnetic Spectrum (EMS) since Gamma Ray towards extended Radio Wave. In fact, Wavelength areas of electro-magnetic emission have diverse names varying from Ultraviolet (UV), Gamma ray, Visible Light, x-ray, Infrared (IR) to Radio Wave [6].

Remote sensing is officially identified as the discipline of finding data in relation to a region, object or event *via* the study of information obtained *via* a tool that is not in direct contact with the region, entity or event in examination [4, 5]. The foundation of *remote sensing* technology is based on the measurement and analysis of the patterns of electromagnetic radiation. Remote sensing uses include a large number of functions, covering the action of satellite arrangements, picture records achievement and memorisation, the consequent data recording, analysis, distribution of saved records and illustration results [5].

Remote detection gives proof that is credible, justifiable, and truthful for decision makers at an assortment of levels [7]. Food security issues, for the most part, happen in remote, horticultural locations with ineffectively developed authority and statistics gathering framework. Checking food production in these regions has necessitated that FEWS NET puts resources into remote estimates that do not depend on reporting of yield and territory planted data of the sort that would be employed. The Famine Early Warning Systems Network, subsidized by the U.S. Organization for International Development, attempts to improve worldwide food protection through the arrangement of significant and early data to guideline makers of people in danger of starvation. Food security, which happens when all individuals consistently have access to adequate, secure, and healthy food to keep up a solid and dynamic life [8], is a basic apprehension in the main survival cultivating economies of arid regions of Africa. Rural inhabitants, depending on rain-fed farming and pastures for their livings, are especially vulnerable to shifts in atmosphere conditions [9]. Monitoring developing conditions utilizing remote sensing data is presently a fraction of early caution that can moderate or even avoid the loss of lives and works related with food security emergencies [10].

Geographic Information System (GIS) is a device that makes visual representations of data and executes spatial investigations with the purpose of making informed decisions. Geographic Information Systems are very useful in being capable to plot and deduct current and coming variations in rainfall, temperature, crop yield, *etc.* [11]. Geographic Information System application in agriculture has been progressively taking part in an essential position in crop production throughout the world by way of assisting farmers in growing production, decreasing costs, and managing their land resources more efficiently [12]. Geographic Information System is applied in agriculture, such as rural mapping performs a crucial function in checking and management of earth and irrigation of any selected arable land. Geographic Information System applied in agricultural mapping acts as imperative equipment for the supervision of the agricultural sector by taking and implementing accurate records into a mapping environment [13]. Geographic Information System is applied in agriculture to help in the management of agricultural resources. Geographic Information System aids in the enhancement of the existing structures of obtaining and generating GIS and resources data [14].

The ability of Geographic Information Systems to study and represent agricultural environments and workflows has found them to be extremely advantageous to those engaged in the agricultural trade. Geographic Information Systems have the ability to examine soil records and conclude which plants should be planted where and how to sustain ground nutrition so that the plants are best advantaged [15]. It assists cultivators to reach higher productions and lower costs through enabling better management of land resources. Agricultural Geographic Information Systems using Geomatics Technology facilitate the farmers to plan and forecast current and future fluctuations in rainfall, temperature, and yield.

REMOTE SENSING AND GEOGRAPHICAL INFORMATION SYSTEM: GENERALITIES

Remote sensing, as an instrument and technique, has developed traditional similarity with other scientific development, for example, the enhancement in an optical lens, electronic detectors and sensors, satellite interfaces, communication structures and the computational processing of recorded data. In the 20th century, the best progress in airborne exploration and picture elucidation was established by the armed forces during the two World Wars [16], and more recently, this novelty was made accessible to the civil residents, leading to its initial appliance in the natural resource supervisions. In the sixties, the National Aeronautics Space Administration started the Television and Infrared Observation Satellite (TIROS-1), which permitted the greatest comprehension of atmospheric circumstances. The expression "remote sensing" was initially employed in the sixties to express any types of monitoring globe from space and launching numerical equipments with the National Aeronautics Space Administration [5]. The recent history of remote sensing of the globe *via* satellite commenced while the Landsat Multispectral Scanner System (MSS) supplied firstly at the beginning of seventies. The Multispectral Scanner System is constituted by multi-spectral bands (4 bands) through sensibly available high spatial resolution (80 m), extensive region (185 × 185 km), and replicating exposure (each 18 days) [17, 18]. Since this period, there have been two thematic Mapper (TM) systems, four supplementary systems of Landsat, and the Enhanced Thematic Mapper Plus (ETM+) in the Landsat series. TM and ETM+ detectors are known as sophisticated multi-spectral searching tools invented to reach elevated picture resolution amid a spatial motion of 30 meters for band 1 to band 5, and band 7, and a spatial motion of 120 meters for band 6 in TM. It is important to note that *band 6* senses thermal (heat) infrared radiation. In addition, The ETM+ has a supplementary band with 15 meters spatial motion [4]. The last Landsat satellite version (Landsat 8) has two main detectors: the Operational Land Imager (OLI) and the Thermal Infrared Sensor (TIRS). These detectors offer seasonal coverage of global landmass at a spatial motion of 30 meters (visible, near infrared and shortwave infrared); 100 meters (thermal); and 15 meters (panchromatic).

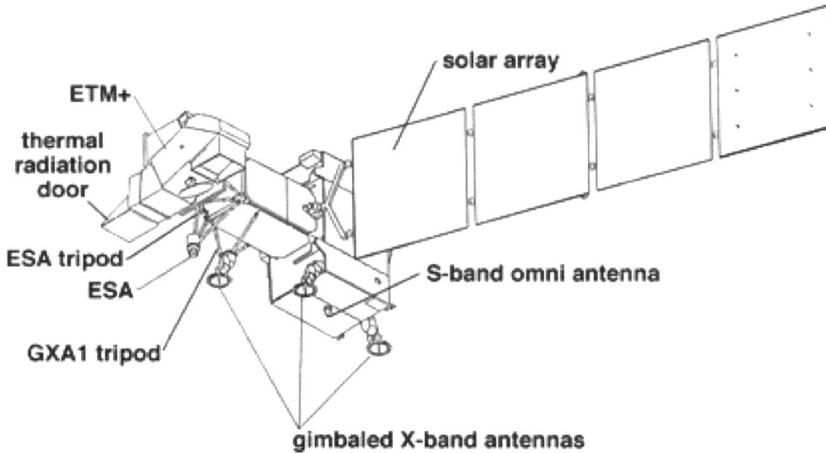

Fig. (1). LANDSAT 7 Satellite composition [19].

Additionally, there have been 5 greater motion SPOT systems, numerous smaller motions AVHRR and GOES systems and National Aeronautics and Space Administration's detector collections on the Earth Observing Systems (EOS), Terra and Aqua satellites in addition to a large assortment of multi-spectral detectors on satellites and aircraft. As well the hyperspectral detectors were expanded and distinguished among an elevated number of bands; these detectors are the European Space Agency's Medium Resolution Imaging Spectrometer (MERIS) and the MODerate Imaging Spectroradiometer (MODIS) [18].

Numerous detectors have been launched and developed in order to attain a greater spatial resolution of 4 meters as the Systeme Probatoire d'Observation de la Terre (SPOT-5), QuickBird, IKONOS, GeoEye-1 and Orbital View-3 (OrbView-3) [4, 10]. Besides, the cited optical remote sensing detectors, as well as novel dynamic remote sensing detectors of short wavelength laser light (LIDAR), long-wavelengths microwaves (RADAR), the Phased Array type L-bans Synthetic Aperture Radar (PALSAR) systems on Japan's Advanced Land Observing Satellite (ALOS) satellite and European Satellite agency's (ESA) Envisat are successfully used. These sensors construct permanent strips of imagery representing wide land regions that match to the platform's flight contour. Recently, the Sentinel project (Sentinel-2A, 2B) launched by the European Space Agency (ESA) and furnished with an optoelectronic multispectral detector for surveys with a motion from 10 to 60 m in the visible, near infrared (VNIR) and short-wave infrared (SWIR) spectral sectors, counting 13 spectral channels, which guarantees the display of dissimilarities in the foliage status, comprising temporal changes, and minimizes the impact on the atmosphere status. The trajectory has an altitude average of 785 km, and the existence of two satellites in the mission allows repeated surveys each 5 days at the equator and each 2 or 3 days at mid-latitudes.

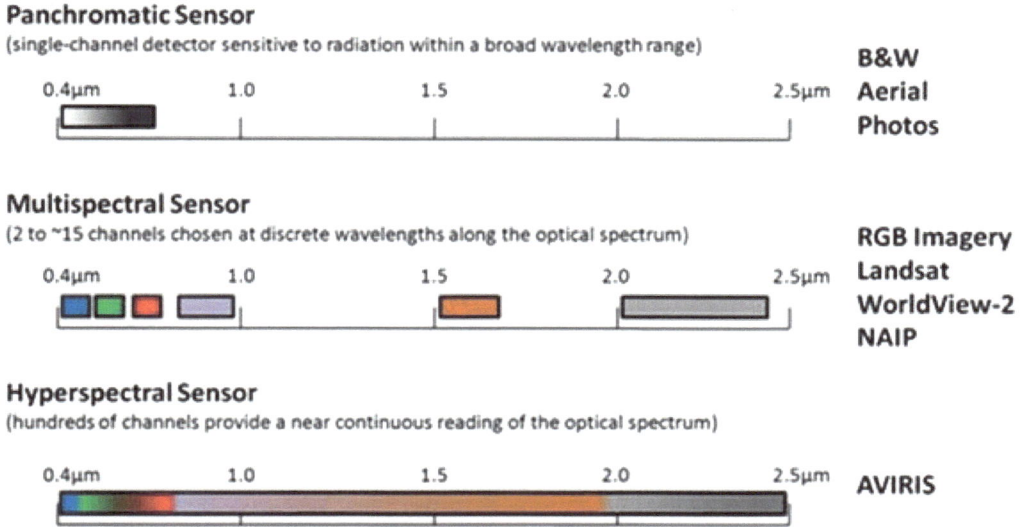

Panchromatic Sensor
(single-channel detector sensitive to radiation within a broad wavelength range)

0.4μm   1.0   1.5   2.0   2.5μm

B&W Aerial Photos

Multispectral Sensor
(2 to ~15 channels chosen at discrete wavelengths along the optical spectrum)

0.4μm   1.0   1.5   2.0   2.5μm

RGB Imagery Landsat WorldView-2 NAIP

Hyperspectral Sensor
(hundreds of channels provide a near continuous reading of the optical spectrum)

0.4μm   1.0   1.5   2.0   2.5μm

AVIRIS

Fig. (2). Spectral resolution of different sensors [20].

Nowadays, there is an enhanced amount of satellite sensor structures being employed to detect and examine the globe by means of a massive volume of data applying an assortment of novel ways for learning changes of earth's ground [17]. There are a lot of different designations of Geographical Information System (GIS) [21], nevertheless, the fundamental notion general to all designations is that GIS is a collection of processes that accumulate, control, operate and symbolize the recorded data amid various categories of spatial elements [22]. This geographical indication information comprises plans and statistics. GIS was initially created in North America in the sixties through the Canada Land Inventory in ecological protection by highlighting the incorporation of spatial recorded data from diverse origins [23]. The progress of digital picture analysis methods and practices from the seventies augmented the link of satellite recorded information in addition to many geographic parameters. Firstly, the incorporation of both methods planned to sustain digital categorization *via* the exploitation of secondary data. In the eighties, the rapid spread of GIS in the arrangement and educational organizations led to concentrating the supervision of spatial records on this skill. At the beginning of the nineties, remote sensing was incorporated with GIS, offering input factors to the spatial platforms. The rising accessibility of high-spatial motion detectors at the commencement of the 21st century has distorted the boundary among knowledge. Remote sensing skill has appeared as a possibly influential device for giving that instruction on natural reserves at different temporal and spatial resolutions.

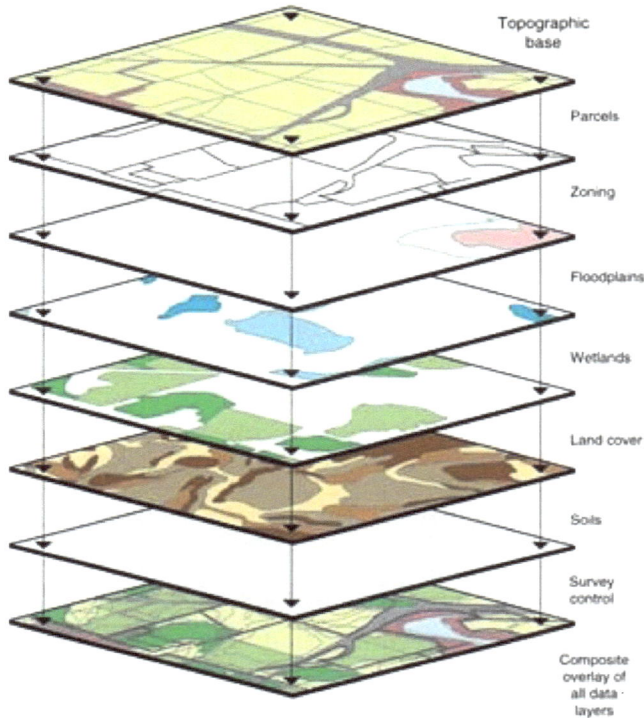

Fig. (3). GISs principle [24].

GIS AND REMOTE SENSING INTEGRATION

Remote sensing, in combination with analogous expansion in GIS, Global Positioning System (GPS) and other land records compilation structures, currently supply a massive quantity of information concerning the ground, in the aim to develop the comprehension of globe systems and better supply for conserving it [5]. GIS is indispensable for getting a further complete vision for remote sensing outcomes of a given domain of attention. Consequently, the assimilation of spatial record has been extremely preferred by the majority of experts. The remote sensing and GIS use has increased much appreciation as ecological resources organization instruments for data compilation and examination [25]. Indeed, GIS and remote sensing are recognized to be not just influential, but in addition economic devices for calculating the spatial organization and developments of ground cover [26 - 30]. Nowadays, there has been a growing importance of measure instruments in the field of remote sensing, GIS and spatial designs [32].

Now, GIS is judged to be an adequate approach that collects geographic details since it is able to combine all foundations of spatial records [5].

The rising complementary between remote sensing and GIS is reasonable, amongst other, for the following causes:

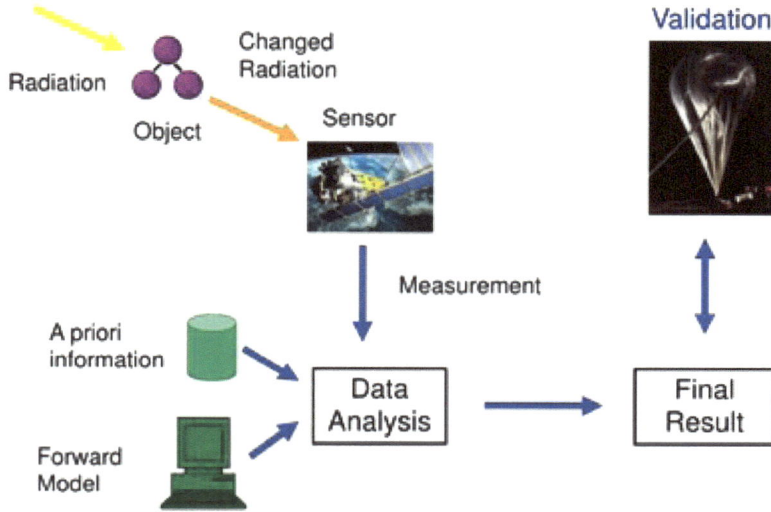

Fig. (4). principles of remote sensing observation [31].

- Together have an understandable regional importance
- They have analogous computer software and hardware requests for data analysis
- Several types of users (scientists, geologists, meteorologists, farmers, *etc.*) frequently apply these tools
- They have comparable study needs, like in the error study case, the configuration of and access to recorded information, the progress of processor, and the explanation approaches.

However, the assimilation of GIS and remote sensing in addition generates some difficulties, principally associated to data accessibility and hardware costs; nevertheless a tendency in the direction of the union of both technologies is clear. A combination of remote sensing data and GIS is facilitated through numerous improvements [33]. The combination of GIS, remote sensing, and statistical designs have permitted the progress of applications in explaining vegetation supervision difficulties at local degrees [34] in addition to landscape and coarser levels [32]. In this logic, the progress of empirical functions against vegetation observation difficulties demonstrated a significant involvement to its sustainable management. Besides, this incorporation indicated diverse scales of progress from peripheral designs, where the addition among the GIS is accomplished to the complete calculation, where the association operates like a uniform system.

SPECTRAL REFLECTANCE OF FOLIAGE

Vegetation is necessary since it offers a fundamental base for all existing organisms. Organizing vegetation *via* remote sensing is helpful since it is capable to establish foliage occurrence and distribution and also how it is affected by ground and atmospheric features. There are few complications in the description of foliage categories on the basis of corresponding spectral responses only, attributable to general heterogeneity of the vegetation coverage category and parameters influencing spectral reactions.

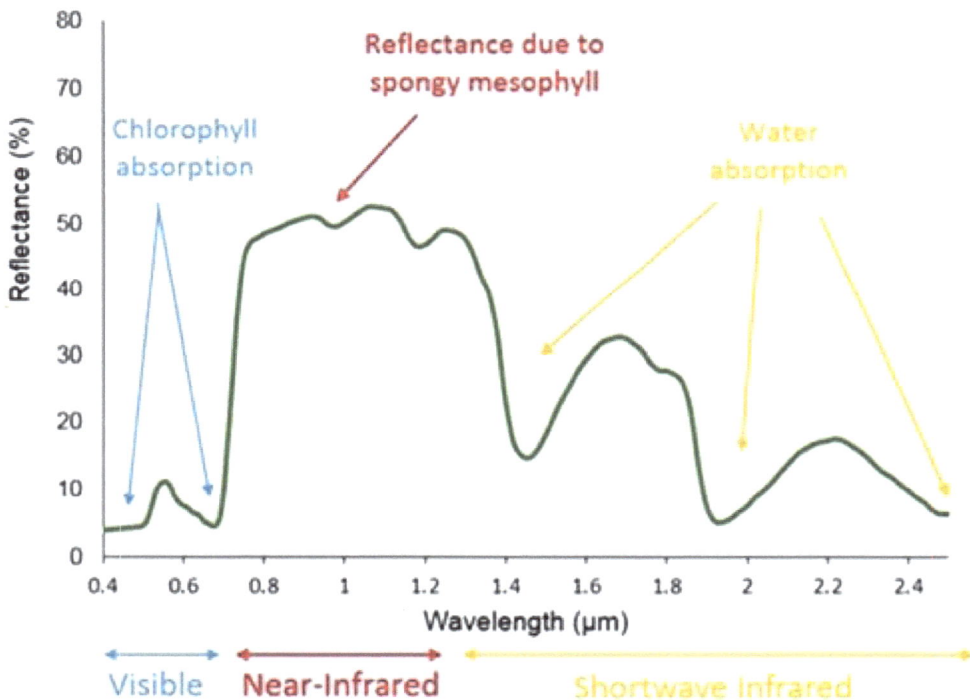

Fig. (5). Vegetation spectral reflectance curves [35].

Furthermore, remote sensing structures propose 4 fundamental constituents to record and to compute available data about a specific region from a given length. These elements comprise the power source, the conduction line, the objective and the satellite detector. Electromagnetic wavelength influences diverse components of trees and plants. These elements consist of stems, leaves, limbs and stalks. The type of return changes, long wavelengths backscatter from stems, but exhibit volume scattering from smaller elements. The density of vegetation canopy will influence the dissemination of wave frequencies. Remote sensing offers significant coverage; mapping and categorization of land cover attributes, for example foliage, ground, and water and plant illustration of spectral reflectance curves for foliage, ground and moisture Fig. (**5**).

Vegetation curve demonstrates small reflectance rates at 0.4 µm, whilst the elevated reflectance values attained approximately the near-infrared and fraction of the mid infrared wavelengths. However, the reflectance range of uncovered ground exhibits reflectance decreasing efficiently among frequency. Its reflectance in the observable waveband is better than the reflectance of vegetation, whilst in part of mid-infrared and near-infrared bands ground reflectance becomes fewer than the reflectance of foliage [36]. The knowledge of surface reflectance attributes supplies an approach which permits the assortment of appropriate wavelength in order to examine the soil surface for a specific task as foliage monitoring.

CHARACTERISTICS OF REMOTELY SENSED RECORDS

The most significant characteristic of remotely sensed information is the resolution. In fact, *resolution* of an image refers to the potential detail provided by the imagery. To measure the spectral response, each satellite has its set of bands corresponding to the wavelength domains. It is important to note that wavelength domains are ranges and each band is strictly defined for a satellite and varies from satellite to satellite. There are 4 categories of resolution for imagery: spatial, temporal, radiometric, and spectral.

Spatial Resolution

Spatial Resolution explains how much detail in a given photographic picture is perceptible to the human eye. The ability to determine or separate details is called spatial resolution. The spatial resolution is linked to the instantaneous field of view (IFOV) of the detector, which indicates the region dimension from which the detector collects the energy at the specific moment [36]. The better is spatial resolution, the more detailed is the picture. In reality, a good spatial resolution decreases the amount of merged pixels, particularly if the site is very divided and ground cover districts have an unequal contour [36]. Images having just visible large features have coarse or low resolution. In opposition, small objects can be detected in high resolution images.

Spectral Resolution

Spectral resolution concerns the capacity of a remote sensing structure to divide the variation in reflectance of the identical land entity at diverse frequencies. The entire energy determined in every spectral detector band is a spectrally-weighted addition of the representation irradiance over the spectral pass band [18]. The utilization of supplementary spectral groups in categorization is advantageous to the accomplishment of greater categorization precision from certain scale.

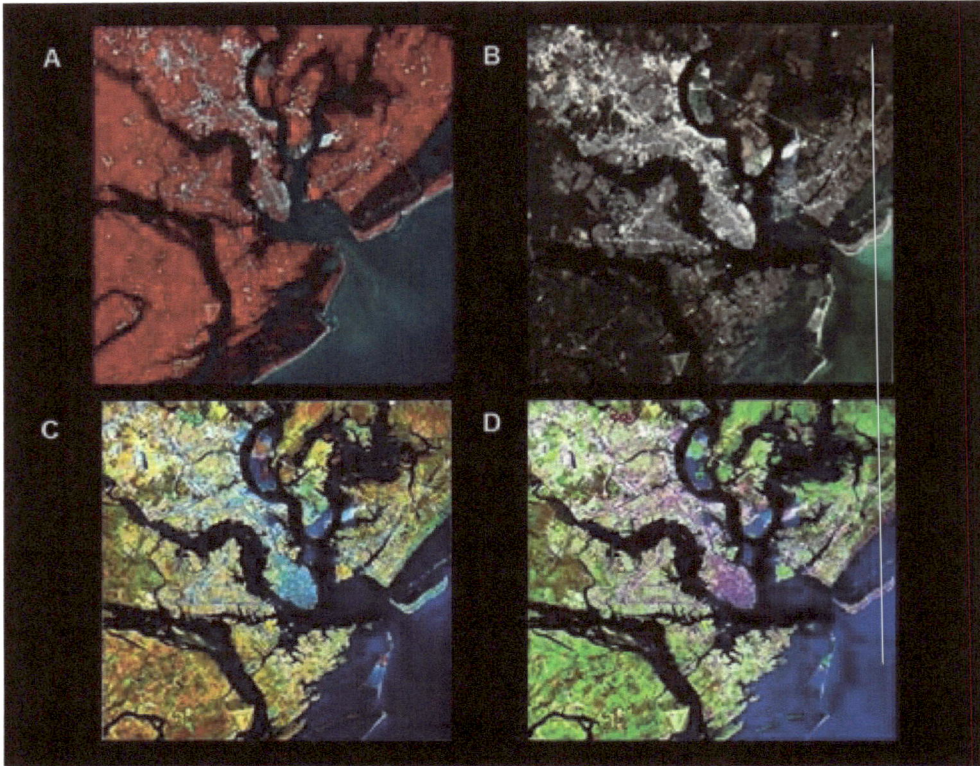

Fig. (6). Fig. (**6 b**) is a real nature composite picture, Fig. (**6 a, c and d**) illustrate two dissimilar false color composites. A false color composite usually improves certain attributes on a picture, attributes that probably not be as evident on a true color composite [39].

Radiometric Resolution

Radiometric resolution concerns the amount of numerical quantization stages employed to evaluate the accumulated data through the detector. Generally, the better the amount of quantization stages the better the radiometric element in the data accumulated by the detector [37]. Radiometric resolution in addition concerns the aptitude of a remote sensing structure to differentiate between the inequalities in the concentration of the radiant power from an objective on the detector.

Temporal Resolution

Temporal resolution concerns the temporal rate of recurrence at which the same soil region is sensed successively by the same sensing structure [37]. A small phase signifies supplementary revolutions per unit of time and is corresponding to an elevated temporal resolution. Satellite temporal resolution fluctuates by means of the latitude of the geographic region being sensed [38].

IMAGE CLASSIFICATION

The purpose of the classification method is to classify all pixels in a given numerical picture into one of the available land cover categories. This classified record can subsequently be employed to construct thematic plans of the territory cover in a representation. Generally, multi-spectral recorded information are employed to achieve the categorization [40]. Image classification is probably the most significant fraction of digital representation. The thematic plot explains the spatial division of identifiable globe surface attributes and offers a detailed explanation over a specified region, rather than a record explanation [18]. Categorization algorithms can be assembled into one of two sorts: parametric and non parametric. Parametric algorithms suppose a specific class of statistical allocation, usually the normal distribution. Non parametric algorithms make no hypothesis concerning the probability distribution and are habitually judged as robust. Hard classification techniques suppose that every pixel symbolizes a uniform district on the soil and explaining just one soil cover category [41]. However in reality, the spatial composition and dissimilarity of ground cover can cause many merged pixels in remotely sensed imagery too in waterless regions, characterized by a low annual pluviometry. There are principally two different methodologies in conventional techniques of classification specifically supervised and unsupervised classification, which are habitually named hard classification. There are soft classification techniques as sub-pixel classification or else blurry classification. In hard classification, 2 types of improbability were discerned. Usually, there are no fixed frontiers amid two ground cover categories. Furthermore, there might be more than one category in one pixel. These difficulties have led to a soft classification notion [42, 43].

Hard Classification

Hard classification employs mathematical processes that try to plot every pixel through allocating it completely to one particular category. The spectrally analogous records will explain thematically comparable entities; and is a principal panorama constituent for every pixel [44]. The conventional hard categorizers employ dual logic to conclude category membership, in that every observation can belong to individual class [45]. As a result of the heterogeneity of soil cover attributes and the restriction in spatial resolution of remote sensing images, merged pixels are regular in coarse- and medium-spatial resolution records. The existence of merged pixels has known trouble that influences the efficient application of remotely sensed informations in per-pixel based categorization [46, 47]. On other hand, deterministic categorization can be more partitioned into four groups: manual classification, supervised classification, unsupervised classification and finally expert knowledge-based classification.

Unsupervised Classification

Unsupervised image classification is the procedure by which every picture in a dataset is recognized to be an element of one of the inherent categories existed in the picture assortment not including the utilization of labelled training examples. Unsupervised categorisation of pictures relies on unsupervised machine learning algorithms for its implementation. Besides, unsupervised classification is an approach through which pixels in a representation are consigned to spectral categories without the operator having an understanding of task existence or else designations of those classes. It is achieved most recurrently by means of clustering techniques. These processes can be employed to establish the amount and position of the spectral classes to choose the spectral category of each pixel [22]. Popular classification algorithms comprise k-means, Hierarchical clustering algorithms and Spectral clustering.

Supervised Classification

The supervised methodology to pixel grouping necessitates the operator to choose representative training information for every of an identified amount of categories. Supervised machine learning is the construction of algorithms that are capable to create general factors and suppositions *via* employing externally supplied instances to forecast the fate of upcoming instances. Supervised machine learning classification algorithms aim at categorizing data from prior information [48]. Classification is carried out very frequently in data science problems. Various successful techniques have been proposed to solve such problems as Rule-based techniques [49], Logic-based techniques [50], Instance-based techniques [51], and stochastic techniques [52].

Supervised classification methods apply also previous understanding of the field, which is greatly useful in obtaining improved classification [42]. Supervised classification supported the concept that recognized information is employed to categorize picture pixels through identifying different training districts of soil cover present in a landscape. Supervised classification is desired by the most part of scientists for the reason that it usually provides more precise categorial descriptions and high precision than unsupervised methods [43]. The statistical classifiers generally used are: minimum distance classifier, the parallelepiped method, and the maximum likelihood algorithm [36].

The maximum likelihood process is a statistical supervised methodology to model identification. The possibility of a pixel belonging to every predefined set of categories is determined, and the pixel is subsequently allocated to the group for which the probability is the maximum [36]. Maximum Likelihood Classifier is one of the most frequent supervised classification methods for parametric data. Besides, maximum likelihood classifier presumes that a pixel has a specific possibility to a specific category. These expectations are equivalent for all groups and the input information in every band is in concordance with the gaussian distribution function [4]. It is imperative to comprehend that the maximum likelihood technique is founded on the theory that the frequency distribution of category bias can be estimated through the multivariate common probability distribution [37]. Since supervised training does not essentially result in category marks that are numerically independent in attribute space, and unsupervised preparation does not automatically result in groups that are important to the scientist, a mixed approach may accomplish both conditions [18].

Expert knowledge-Cased Classification

Expert classification software offers a rules-based methodology to multi-spectral picture organization, post-classification enhancement, and GIS designing. In real meaning, an expert classification scheme is a chain of regulations, also called decision tree, which illustrates the situations under which a set of little stage component instruction gets abstracted into a set of elevated stage informational groups [53]. In fact, decision trees are constructed *via* a heuristic classification called recursive partitioning. This procedure is also generally identified as a dividing method because it splits the records into subsets, which are subsequently split frequently into even smaller subsets, until the process stops when the algorithm verifies the records within the subsets are sufficiently homogenous, or another stopping criterion has been occured [54].

Soft Classification

Soft classification algorithms are created to handle, with the complexity of "merged pixels", *via* depicting spatially the varied disposition of ground cover with regard to wide surfaces. This classification theory is an option to the usual "hard" classifier and affording the analyst by means of a level measure which the specified pixel belongs to a few or of the entire selected categories, and delivers to the researcher the result as to which group the pixel has to be categorized [55]. Fuzzy set statement, proposed by Zadeh [56], has a function in accordance with indecision in GIS and remote sensing. Soft classification can be further instructive at the limits among land cover objects, and gives a more useful and potentially more precise choice when compared with hard classification [43]. The fuzzy classifier, called also soft classifier, does not allocate every picture pixel to a distinct category in a clear tendency. As an alternative, every pixel is specified with bias significance for each class. Bias status varies with a value between 0 and 1, and offers a level measurement to which the pixel fits or resembles to a particular class, only as parts or fractions employed in linear combination design [37]. Several studies have tried to reorganize the information from merged pixels [56 - 58]. Baatz *et al.* [59], build algorithms on the basis of basic mathematics. The numerical procedure of fuzzy logic consists to substitute the strict coherent notice 0 or 1 through a continuous interval between 0 and 1, where 0 indicates strictly No and 1 indicates strictly Yes [59].

Object Criented Image Analysis

Lately, an additional form of classifier identified as object-based, has appeared, and has usually improved, through short band, important spatial resolution records [60]. Utilization of a classified categorization, established on an object oriented theory, has numerous benefits as the replacement for the pixel-driven procedure. Image objects, in addition to spectral information, include supplementary characteristics that can be employed for categorization objectives [61]. In this case, the categorization method starts by means of a division of adjacent pixels inside uniform objects or units [62].

Image segmentation consigns to the separation mode of an input picture into spatially isolated and adjacent regions or segments. These sections are patches consisted of identical pixels. These pixels allocate a greater internal spectral uniformity amid themselves than external uniformity with pixels in other areas [63]. Image division and segmentation may be shown in bottom-up or else top-down approaches or else a dual combination. In the top-down process, the input image is divided into many harmonized sections. However, in the bottom-up process, pixels are connected jointly to compose sections that are subsequently combined [38].

LAND USE AND LAND COVER

Definitions

The difference between land cover and land use is simple to make at a theoretical stage. Land cover is the viewed material territory, as seen from the soil or *via* remote sensing, as well as the spontaneous or cultivated vegetation, or else to constructed edifices such as roads, buildings, *etc.* which occupy the surface of the globe. Ice, water, uncovered rocks or else sand areas are considered as land cover

[64]. Land utilization is founded upon function, the reason for which the ground is being employed. Consequently, land utilization can be identified as a sequence of actions undertaken to make one or more services. A specified land use could take position on one or more portions of land and thus many land utilizations may happen on a similar land portion [64].

Vegetation Observing and Mapping

The properties and characteristics of foliage are considered as essential traits of landscapes. The nature of the foliage in a region is established through a mixture of results linked to soils, climate, fire, history and Man controls. Vegetation and foliage mapping has an extensive history which consists of an assortment of situations and a large variety of geographic degrees. Plots of vegetation try to verify what the foliage sort would be in the nonexistence of human controls. Records of real flora try to differentiate the plants as it is in a specific district [65]. Observing of foliage change *via* remote sensing is supplying a better apprehension of the foliage vigour and state in addition to values of change of natural foliage to different land utilizations. Primary foliage maps and vegetation imageries made *via* remote sensing were founded on the ocular explanation of aerial pictures [4].

Precise and reliable data on cover area transformation is imperative. Cover area transformation attributes frequently provide information on natural activity. The main theory of precision assessment is that it evaluates the mapped soil categorization to higher quality reference records, collected *via* a sample based approach [66]. The higher quality reference information can be acquired by soil collected records; nevertheless as this is costly and labor intense it is more usually acquired *via* satellite imagery or aerial photography with finer spatial resolution than data that was employed to make map data. When relying on imagery for reference data and there is no elevated resolution imagery obtainable, higher quality records can be collected by a process considered as more precise, such as a human explanation of reference records [66].

With the initiation of Landsat project in the seventies, there was particular attention in the capacity for mapping foliage of wide regions in a more efficient approach than conventional air photo explanation. Dynamic studies and numerous ideas are employed for using digitized satellites pictures for mapping foliage. They comprise classification *via* per-pixel classifiers; by means of contextual or else spatial information in the categorization procedure [67, 68], and by division of images into polygons in a phase without image categorization and classification. Furthermore, there are three main features that are concentrated whilst examining natural locations and land cover modifications. Those comprise the amount and the degree of the transformation, the kind of the transformation, and the spatial model of the transformation estimating the spatial allocation and transformation liaison [69].

In the examination of ground exploitation and ground cover transformation; it is initially indispensable to visualize the implication of transformation in order to distinguish its authentic states. Nevertheless, in land cover and use, the implication and visualization of transformation are greatly significant. In land cover transformation, it is essential to differentiate among two sorts of transformation: change and adaptation [70]. Land cover adaptation engages a modification from a cover sort to another. Land cover change engages variations of composition or role not including a general switch from one sort to another. Several parameters affect foliage state and vigour, varying from water deficiency and diseases to acid precipitation and atmosphere pollution. Remote sensing proposes an optional theory whose power is spatial coverage, when combined with soil sub-region of soil testing demonstrate to be particularly useful for supervising vegetation strength.

Ways of Assessing Foliage State

Multi-Temporal Analysis

Multi-temporal satellite is generally employed in this objective. In which, obtained pictures in diverse dates to the identical position are co-recorded in the aim to evaluate the spectral rates and also to supervise the foliage vigour. In numerous situations, the examination of multitemporal image representation has attested as efficient for observing defoliation of vegetations caused by pests [71].

Macroscopic Visualisation of Vegetation Conversion and Change

A different type of variation in foliage and vegetation of important significance is the conversion of foliage and vegetation sorts. The most noticeable case of this type of transformation is erosion, which is considered as one of the most considerable types of soil use transformation stirring on the globe. Ground erosion leads to important on- and off-site impacts such as an important decline in the productive capability of land and sedimentation. The principal aspects influencing the quantity of soil erosion mainly rely on foliage cover, geography, soil nature, and climate. Thus, several innovative Earth observation approaches will be examined for their potential and impact on observing ground characteristics and corresponding ground erosion phenomena. Remote sensing proposes a unique opportunity to map, monitor, quantify, and analyze, in detail, the processes that contribute to soil loss as a result of water erosion [72].

APPLICATION OF REMOTE SENSING IN VEGETATION INVENTORY

Sampling can be described as the procedure of acquiring information through measuring just a fraction of land and describing inference for the entire region. Where spatial data is required, remote sensing proposes appropriate techniques and can be noticeably enhanced by employing it for land exploitation records [73]. Local estimations necessitate different approaches from local or regional measurements. Therefore, remote sensing has begun to be a well-known instrument for multi-scale evaluation of vegetal resources [32]. Normalized Difference Vegetation Index (NDVI) quantifies foliage through calculating the variation involving near-infrared (which foliage strongly reflects) and red light (which foliage absorbs) [74]. EVI is analogous to Normalized Difference Vegetation Index (NDVI) and can be employed to measure foliage greenness [75]. Nevertheless, EVI corrects for some atmospheric conditions and canopy background noise and is more sensitive in regions with intense foliage. In addition, the SAVI is structured comparable to the NDVI but with the addition of a soil brightness correction factor [76].

Besides, the MERIS Global vegetation index is an estimation of the existence of healthy live green foliage. It was optimised to be strong in atmospheric conditions and surface reflectance and constrained to provide a fraction estimation of absorbed photosynthetically active radiation [77]. The Medium Resolution Imaging Spectrometer (MERIS) Terrestrial Chlorophyll Index (MTCI) offers information on the foliage chlorophyll content (amount of chlorophyll per unit area of soil) [78]. Moreover, the NDRE *(Normalized Difference Red Edge)* is an indicator that can just be formulated when the Red edge band is accessible in the sensor. It is receptive to chlorophyll quantity in leaves (how green appears in a leaf), variability in leaf surface, and soil setting effects [79].

On the other side, the utilization of GIS tools in vegetation organization and other domains of natural resource managing have amplified continuously [80]. This high technology has permitted the supervision and incorporation of a significant amount of temporal and spatial data [80]. Detectors collect remotely sensed information from lands and these are treated and deducted to extract data, of which there are 3 stages of aspect. The first stage assigns to data on the spatial degree of vegetal foliage, which is capable to be employed to calculate foliage dynamics, the second stage includes genus characteristics within encircling cultivated districts and areas and the last grade presents data on the biophysical attributes of covered lands [81]. The multiplicity of sensor scheme devices actually accessible offers data with broad to good spectral imagery, from high to low spatial imagery quality and other features appropriate for qualitative and quantitative examination of covers [82]. A moderately novel optical detector category and the imaging spectrometer symbolize a scientific progress in remote sensing, as they evaluate the returned indication with a well spectral imagery quality [83]. Applications of GISs are considered as one of the most important scientific progresses for natural resource measurement through the 20th century. This method authorized farmers to renew information in a pertinent way. Remote sensing offers quick treatment of wide zones, permanent records, and effectiveness in period and contact to unattainable zones [84]. While a lot of appliances of GIS in agriculture have been planned with broad spatial levels, Pernar and Storga [85] have indicated that GIS can help supervision and mapping for regions less than 1 ha, assisting to resolve problems (pesticide managements, replanting, *etc.*) in every design independently. A good employ of remote sensing in agriculture necessitates comprehension of the spectral characteristics of the concerned vegetation constituents. The relations of electromagnetic radiation among the elements of a cover have to be taken under consideration on the microscopical and macroscopical levels [86, 87]. The satellite picture representation is an indispensable constituent in the amelioration of novel equipment for agriculture management. The resource data required for culture is principally needed for two kinds of management and scheduling actions, specifically calculated (area, health, growth) and operational scheduling (harvesting treatments) [88]. Data extraction from remote sensing records can be explained in terms of designs of the detector, environment, and the landscape [89]. In remote sensing of cultures, mapping of the feature 'culture type' (as age class, tree species, *etc.*) supposes that the picture pixel is totally enveloped by only one vegetation sort and therefore, several pixels compose a culture stand. In this circumstance, picture categorization would be an appropriate processing algorithm. Besides, Franklin [90] has declared common stages for forecast or evaluation of vegetation factors:

a. Create field observation positions in a culture region,
b. Accumulate culture state data of positions,
c. Obtain picture descriptions of the positions,
d. Situate the positions on the picture,
e. Extract the remote sensing records from the locations,
f. Create a prototype describing the ground and spectral records,
g. Apply the prototype to forecast vegetation factors for all vegetation pixels on the basis of spectral records. The remote sensing records and data for culture and natural resources can accumulated through 3 techniques: field measuring, aerial measuring in addition to satellites tools.

Field Measuring

The combination of plant field observations and data derived from satellite image categorization makes it feasible in order to construct statistical models which can be used to establish the probability of a vegetation type existing in a pixel of recognized class. These models are used to build an indirect connection between vegetation characters and spectral signatures which give an indication of the habitat conditions through an overall image categorization. In fact, one of the major constraints of remote sensing devices on satellites or airliner is the cover confuses the trunks and tree branches when versioned from over. This has been an essential restriction to their employ more commonly in estimation. Nevertheless, tools which are placed on the soil are currently becoming accessible to permit remote estimation of the good aspect of plants [91].

Furthermore, proximal sensing induces light-weight transportable spectrometers for the NIR, SWIR, and TIR wavelength intervals for mobile mapping uses. The field spectrometers are ideal tools for fast *in situ* spectral estimations or a proficient land accuracy for remote sensing uses. SphereOptics proposes modern hyperspectral results in the grounds of proximal sensing from 0.35×10^{-6}m to 15×10^{-6}m, which are recognized throughout environmental discipline [92].The GreenSeeker system utilizes optical detectors to evaluate and calculate crop variability. It subsequently builds a specific recommendation to manage the variability of crop [93].

Moreover, Nitrogen is a necessary constituent for plant development. Previous research [94] shows that use of remote sensing to calculate N in wheat canopies is efficient, and estimation of Canopy Chlorophyll Content Index (CCCI) from canopy spectral reflectance of wheat showed a good relationship with N uptake (kg N/ha). Estimation of leaf superficies index is the main support factor, helpful to verify how much daylight a stand takes and, therefore, what the photosynthetic generation of support might be. Fournier *et al.* [95] have established a scheme to determine leaf size indicator from the soil unless felling of plants by means of a device taking into consideration the action of rays of daylight, pending from whichever element in the atmosphere above, like 'indicators' which are being screened through the vegetal cover. On the one hand, the leaf area index (LAI) is not just a significant factor utilized to explain the geometry of vegetation cover but also a main input variable for environmental designs [96]. One of the most generally utilized processes for LAI calculation is to create an empirical association between the LAI and the vegetation indicator. Besides, the LAI is strongly related to the vegetation respiration, transpiration, and photosynthesis. The development dynamics and seasonal evolution of vegetation can also be timely reflected by the LAI [97]. Another idea for calculating leaf size index from soil is to capture a cover preview, generally by means of a large-angle lens, exploring vertically and spatially from the soil below [91].

Aerial Measuring

This technique applies tools taken on satellites or aeroplane in the aim to study a zone composed by several hectares of plants, furnishing data valuable for different reasons. This aeroplane manoeuvre changes greatly in elevation and speed. Examples of these flying machines are aerial imagery, spectrometry in addition to laser inspection [91].

The latest and empirical techniques (soil-based and remote sensing devices) employ the signal of laser radiance to build a 3D representation of plants in a given zone. In the case of remote sensing, laser calculation is frequently named 'LIDAR', a short form for LIght Detection And Ranging; similar to the most generally identified 'radar', a short form for RAdio Detection And Ranging, which employs radio, instead light signals [91]. These soil-based and laser calculation devices are visibly showing the capacity for specified characterization of plant features. Nevertheless, they will necessitate significant research progress before their employ in training for broad-scale in plant lists [91].

Aerial photographic analysis has been the most commonly utilized type of remote sensing for ecological appliances actually, with procedures being well established. This wide utilization stems from the accessibility of aerial photographs, and the fact that they can give very high spatial resolution records, down to metre precision, are not historically obtainable from other sensors. For both aerial photographs and space-borne remotely sensed records, the scale of the picture will resolve the potential of the analysis in a large extent. Unmanned Aerial Vehicles and Systems (UAV or UAS) have become progressively more accepted in recent years for agricultural research purposes. UAS are able to acquire pictures with high spatial and temporal resolutions that are perfect for applications in agriculture [98]. Aerial measure presents many advantages as a flexible availability, a relative low-cost and a very high spatial resolution. However, this technique of visualisation has some disadvantages: it may be limited in height above ground, a geographic distortion and finally it may require certification to operate [99].

Satellites

Satellites propose one of the most inclusive structures of remotely sensed and descriptive data from vegetation. Satellites detector equipments may manage radiation returned from the globe surface; in addition of others dynamic release signals of radar microwave. For example, the Landsat satellite reference is employed extensively for vegetation objectives. Landsat satellites comprise globe examination structures, for instance the Nippon ALOS, the American IKONOS, the Indian IRS, the American NOAA-AVHRR, American Quickbird, French

SPOT and finally the Nippon JERS [91]. The advantages are numerous: stable and clear images, good historical data and finally the large area within each image. The disadvantages are the elevated cost for high spatial resolution images, data may not be assembled at decisive times and also the clouds may hide ground features. The following Table **1** resumes the differences of various platforms

Table 1. Advantages and disadvantages of different platforms.

Platform	Advantages	Disavantages
Field or ground Method	• Flexible availability during day and night • Helpful for real-time spraying applications	
UAV	• moderately low cost • elevated spatial resolution • variable sensors	• Geographic distortion
Manned aircraft	• moderately high spatial resolution • Changeable sensors • Flexible availability	• High fixed costs • Availability depends on weather conditions
Satellite	• Free mesospatial image availability • Stable images • Good historical data	• High cost for high spatial resolution images • Fixed schedule (except for SPOT4) • Data may not be collected at critical times

Synchronized Applications of Modelisation and Remote Sensing in Vegetation Analysis

Designing and simulation are indispensable for the investigation of composite active structures, for the reason that they are characteristics for environmental segments [73]. There are principally three distinct theories employed in plant listing *via* remote sensing in the aim to measure biophysical factors from spectral signs given *via* satellite pictures (1) physical (2) empirical, and (3) a dual mixture (as neural networks) [100].

Empirical Theory

It is normally applied in local or regional land inventories. The vegetation of interest and the corresponding characteristics are determined on the basis of mathematical connections acquired *via* accumulating training records on the spectral marks of an assortment of objects. These techniques necessitate big sets of stable soil accuracy records.

Physical Theory

It is relying upon a comprehension of physical rules managing the effect of sun radiation in foliage cover, which are defined scientifically by reflectance designs. The reflectance designs construct connections involving the biophysical characteristics of foliage cover and spectral sign. In opposition to empirical approaches, physically-based theories have the benefit that they can take into consideration results of changing cover configuration, covering form, and ground that all control the radiation calculations. Spectral combination examination employs easy reflectance designs to calculate approximately the fractional composition of the main kinds of objects contained by figure pixels [101].

Dual Mixture: Artificial Neural Networks

The investigation of recent efforts on the appliance of Artificial Neural Networks (ANN) has shown that neural networks are principally employed for investigating and identifying objectives that are affiliated to the flexibility class [102]. ANN represented a scientific and useful groundwork in resolving a variety of difficulties based on smart processing of Earth remote sensing pictures (panchromatic, multispectral, coloured) search for unbending objects and regions of interest by means of the expanded spectrographic method and the generalized metric [103]. Several studies [104, 105] are related to the use of modern Convolutional Neural Networks (CNN) for processing panoramic full-colour globe remote sensing pictures obtained *via* unmanned aerial vehicles (UAVs); some methods and devices to develop their effectiveness and performance through the exploration and detection of the objects among the essential completeness and precision, which still stays unanswered even with a profusion of programs, are developed [106]. The current formulation of the mission of discovering and identifying a target (object, region,*etc.*) *via* a neural network comprises the actions of selecting the type, setting the factors of the ANN and arranging the input records. The multi-category and single-category difficulties were judged as learning fraction [107 - 110]. The primary assignment is thought of as the exploration and detection of objects of numerous categories concurrently. The following assignment engages the exploration *via* an artificial neural network of objects of a single category.

CONCLUSION

Remote sensing offers an information foundation for making baseline records on natural reserves, a condition for preparing and execution, and checking of every development of the environmental plan. Global Positioning System (GPS) proposes a perfect setting for the incorporation of spatial records and characteristic statistics on natural sources for creating the geographical plan of a region taking into consideration the economic and cultural requirements of vegetation. The digital model created from the information prepared by Global

Positioning System by digital photogrammetric process facilitates further enlightening the geographical plans.

In spite of remarkable progress in the captor equipment, record processing and investigation/explanation methods, some particular inputs such as, growth of GIS-based land evaluation designs for land capacity, land irrigability, appropriateness of land for precise use, are not assured. Objective impact evaluation by means of space expertise, progress of ecological designs to plan future situation, *etc.* could not be addressed. Hyperspectral data and high spatial resolution data may facilitate generating most favourable land employ plan or exploitation plan for sustainable development of territory and irrigation resources.

CONSENT FOR PUBLICATION

Not applicable.

CONFLICT OF INTEREST

The author(s) confirms that there is no conflict of interest.

ACKNOWLEDGEMENTS

Declared none.

REFERENCES

[1]   Liaghat S, Balasundram SK. A review: The role of remote sensing in precision agriculture. Am J Agric Biol Sci 2010; 5(1): 50-5.
[http://dx.doi.org/10.3844/ajabssp.2010.50.55]

[2]   Usha K, Singh B. Potential applications of remote sensing in horticulture—A review. Sci Hortic (Amsterdam) 2013; 153: 71-83.
[http://dx.doi.org/10.1016/j.scienta.2013.01.008]

[3]   Lindblom J, Lundström C, Ljung M, Jonsson A. Promoting sustainable intensification in precision agriculture: review of decision support systems development and strategies. Precis Agric 2017; 18(3): 309-31.
[http://dx.doi.org/10.1007/s11119-016-9491-4]

[4]   Lillesand MT. KIefer WR, Chipman NJ Remote sensing and image interpretation. 6th ed., New York, USA: John Wiley and Sons, Inc. 2008.

[5]   Chuvieco E, Huete A. Fundamental of satellite remote sensing. New York, USA: Taylor and Francis Group 2010.

[6]   Stellingwerf DA, Hussin YA. Measurements and estimations of forest stand parameters using remote sensing. Zeist, The Netherlands: VSP BV 1997.

[7]   Pielke R. When the numbers don't add up. Nature 2007; 447(7140): 35-7.
[http://dx.doi.org/10.1038/447035a]

[8]   Ross KW, Brown ME, Verdin JP, Underwood LW. Review of FEWS NET biophysical monitoring

requirements. Environ Res Lett 2009; 4(2): 024009.
[http://dx.doi.org/10.1088/1748-9326/4/2/024009]

[9]   Verdin J, Funk C, Senay G, Choularton R. Climate science and famine early warning. Philos Trans R
        Soc Lond B Biol Sci 2005; 360(1463): 2155-68.
        [http://dx.doi.org/10.1098/rstb.2005.1754] [PMID: 16433101]

[10]   Tilman D, Cassman KG, Matson PA, Naylor R, Polasky S. Agricultural sustainability and intensive
        production practices. Nature 2002; 418(6898): 671-7.
        [http://dx.doi.org/10.1038/nature01014] [PMID: 12167873]

[11]   Alqurashi A, Kumar L. Investigating the use of remote sensing and GIS techniques to detect land use
        and land cover change: A review. Adv Remote Sens 2013.
        [http://dx.doi.org/10.4236/ars.2013.22022]

[12]   Sharma R, Kamble SS, Gunasekaran A. Big GIS analytics framework for agriculture supply chains: A
        literature review identifying the current trends and future perspectives. Comput Electron Agric 2018;
        155: 103-20.
        [http://dx.doi.org/10.1016/j.compag.2018.10.001]

[13]   Montgomery B, Dragićević S, Dujmović J, Schmidt M. A GIS-based Logic Scoring of Preference
        method for evaluation of land capability and suitability for agriculture. Comput Electron Agric 2016;
        124: 340-53.
        [http://dx.doi.org/10.1016/j.compag.2016.04.013]

[14]   Swain KC, Chiranjit S. Mapping of agriculture farms using GPS and GIS technique for precision
        farming. Int J Agric Eng 2018; 11(2): 269-75.
        [http://dx.doi.org/10.15740/HAS/IJAE/11.2/269-275]

[15]   Rozenstein O, Karnieli A. Comparison of methods for land-use classification incorporating remote
        sensing and GIS inputs. Appl Geogr 2011; 31(2): 533-44.
        [http://dx.doi.org/10.1016/j.apgeog.2010.11.006]

[16]   Khorram S, van der Wiele CF, Koc FH, Nelson SA, Potts MD. Remote sensing: past and
        present.Principles of Applied Remote Sensing. Cham: Springer 2016; pp. 1-20.
        [http://dx.doi.org/10.1007/978-3-319-22560-9_1]

[17]   Sabins FF. Remote sensing, principles and interpretation. 3rd ed., New York, USA: Freeman WH and
        Company 1997.

[18]   Schowengerdt RA. Remote sensing, models and methods for image processing. 3rd ed., New York,
        USA: Elsevier Inc 2007.

[19]   NASA.   Components   of   LANDSAT   7   Retrieved   December   10,   2019,   from
        https://airbornescience.nasa.gov/sge/landsat/l7d.html

[20]   Zuly Z. The Electromagnetic Spectrum Retrieved December 3, 2019, from
        www.zulyzami.com/The+Electromagnetic+Spectrum

[21]   Clarke KC. Getting started with geographic information systems. 3rd ed., New Jersey, USA: Prentice
        Hall 2001.

[22]   Richards JA, Jia X. (4th edition): An Introduction. Berlin, Heidelberg, Verlag, 2006 Remote sensing
        digital image analysis An Introduction. 4th ed., Berlin, Heidelberg : Verlag 2006.

[23]   ERDAS. Field Guide, Atlanta, USA, leica geosystems, GIS and mapping 2003.

[24]   Jensen JR, Jensen RR. Introductory geographic information systems. New Jersey, USA: Pearson
        Higher Edition 2012.

[25]   Rindfuss R, Entwisle B, Walsh SJ, *et al.* Continuous and discrete: where they have met in Nang Rong,
        Thailand.Linking people, place and policy: A GIS science approach. Boston, USA: Kluwer Academic
        Press 2002.
        [http://dx.doi.org/10.1007/978-1-4615-0985-1_2]

[26] Tottrup C, Rasmussen MS. Mapping long-term changes in savannah crop productivity in Senegal through trend analysis of time series of remote sensing data. Agric Ecosyst Environ 2004; 103(3): 545-60.
[http://dx.doi.org/10.1016/j.agee.2003.11.009]

[27] Giridhar MV, Viswanadh GK. Evaluation of watershed parameters using RS and GIS. Proceedings of the 11th Aerospace Division International Conference on Engineering, Science, Construction, and Operations in Challenging Environments. USA. 2008.
[http://dx.doi.org/10.1061/40988(323)124]

[28] Wilson JP, Fotheringham SA. The hand book of geographic information science. Victoria, Australia: Blackwell publishing 2008.

[29] Zhiliang C, Xulong L, Xiaochun P, Zhencheng X, Zhifeng W. Land use/cover changes between 1990 and 2000 based on remote sensing and GIS in Pearl River Delta, China. Proceedings of SPIE.

[30] Dewan AM, Yamaguchi Y. Land use and land cover change in greater Dhaka, Bangladesh: Using remote sensing to promote sustainable urbanization. Appl Geogr 2009; 29(3): 390-401.
[http://dx.doi.org/10.1016/j.apgeog.2008.12.005]

[31] A Richter. Satellite Remote Sensing of Tropospheric Composition – principles, results, and challenges. EPJ Web of Conferences 2010; 9: 9-181.

[32] Franklin SE. Remote sensing for sustainable forest management. Boca Raton, USA: Lewis Publishers 2001.
[http://dx.doi.org/10.1201/9781420032857]

[33] Congalton RG. A comparison of sampling schemes used in generating error matrices for assessing the accuracy of maps generated from remotely sensed data. Photogramm Eng Remote Sensing 1988; 54: 593-600.

[34] Murray AT, Snyder S. Spatial modeling in forest management and natural resource planning. For Sci 2000; 46: 153-6.

[35] Vegetation Spectral Signature Cheat Sheet (accessed on May 20 2020) https://grindgis.com/remote-sensing/vegetation-spectral-signature-cheat-sheet2020.

[36] Mather PM, Brandt T. Classification methods for remotely sensed data . 2nd editionRatonn Boca. CRC Press, Taylor and Francis Group 2009.
[http://dx.doi.org/10.1201/9781420090741]

[37] Mather PM. Computer processing of remotely sensed images. 3rd ed., Chichester, West Sussex, UK: John Wiley and Sons, Ltd 2004.

[38] Gao J. Digital analysis of remotely sensed imagery. New York, USA: The McGraw-Hill Companies Inc 2009.

[39] http://employees.oneonta.edu/baumanpr/geosat2/RS-Introduction/RS-Introduction.html

[40] Lillesand MT, Kiefer WR. Remote sensing and image interpretation. (5th edition).. New York, USA: John Wiley and Sons Inc 2004.

[41] Markham BL, Townsend RG. Land cover classification accuracy as a function of sensor spatial resolution. Proceedings of the 15th International Remote Sensing of the Environment. 11–15 May.; Ann Arbor, MI, USA: University of Michigan Press 1981; pp. 1075-90.

[42] Key J, Yang P, Baum B, Nasiri S. Parameterization of shortwave ice cloud optical properties for various particle habits. J Geophys Res 2002; 107: 4181-91.
[http://dx.doi.org/10.1029/2001JD000742]

[43] Foody GM, Atkinson PM. Uncertainty in remote sensing and GIS. London, UK: John Wiley and Sons 2002.
[http://dx.doi.org/10.1002/0470035269]

[44]   Jensen JR. Introductory digital Image processing- A remote sensing perspective. 3rd ed., Upper Saddle River, NJ, USA: Prentice Hall 2004.

[45]   Foody GM. Image classification with a neural network: From completely-crisp to fully-fuzzy situations. In: PM Atkinson, NJ Tate, Eds. Advances in remote sensing and GIS analysis. Chichester: Wiley 1999; pp. 17-37.

[46]   Fisher PF. Animation of reliability in computer-generated dot maps and elevation models. Cart Geograph Inf Sys 1997; 23(4): 196-205.

[47]   Hu X, Weng Q. Estimation of impervious surfaces of Beijing, China, with spectral normalized images using LSMA and ANN. Geocarto Int 2010; 25(3): 231-53.
[http://dx.doi.org/10.1080/10106040903078838]

[48]   Persello C, Bruzzone L. Active learning for domain adaptation in the supervised classification of remote sensing images. IEEE Trans Geosci Remote Sens 2012; 50(11): 4468-83.
[http://dx.doi.org/10.1109/TGRS.2012.2192740]

[49]   Li M, Zang S, Zhang B, Li S, Wu C. A review of remote sensing image classification techniques: The role of spatio-contextual information. Eur J Remote Sens 2014; 47(1): 389-411.
[http://dx.doi.org/10.5721/EuJRS20144723]

[50]   Weng Q. Remote sensing of impervious surfaces in the urban areas: Requirements, methods, and trends. Remote Sens Environ 2012; 117: 34-49.
[http://dx.doi.org/10.1016/j.rse.2011.02.030]

[51]   Cheng G, Han J. A survey on object detection in optical remote sensing images. ISPRS J Photogramm Remote Sens 2016; 117: 11-28.
[http://dx.doi.org/10.1016/j.isprsjprs.2016.03.014]

[52]   Han D, Mo Y, Wu J, Weerakkody S, Sinopoli B, Shi L. Stochastic event-triggered sensor schedule for remote state estimation. IEEE Trans Automat Contr 2015; 60(10): 2661-75.
[http://dx.doi.org/10.1109/TAC.2015.2406975]

[53]   Elnaggar AA, Noller JS. Application of remote-sensing data and decision-tree analysis to mapping salt-affected soils over large areas. Remote Sens 2010; 2(1): 151-65.
[http://dx.doi.org/10.3390/rs2010151]

[54]   Mustafa AA, Singh M, Sahoo RN, *et al.* Land suitability analysis for different crops: a multi criteria decision making approach using remote sensing and GIS. Researcher 2011; 3(12): 61-84.

[55]   Jensen JR. Introductory digital image processing: A remote sensing perspective. 2nd ed., New Jersey, USA: Prentice Hall 1996.

[56]   Zadeh LA. Fuzzy sets. Inf Control 1965; 8: 338-53.
[http://dx.doi.org/10.1016/S0019-9958(65)90241-X]

[57]   Foody GM. Fully fuzzy supervised image classification: Remote sensing in action. Nottingham: Remote Sens Soc 1995; pp. 1187-94.

[58]   Maselli F, Conese C, Filippis D, Norcini S. Estimation of forest parameters through fuzzy classification of TM data. IEEE Trans Geosci Remote Sens 1995; 33(1): 77-84.
[http://dx.doi.org/10.1109/36.368220]

[59]   Baatz M, Heynen M, Hofmann P, *et al.* eCognition user guide. München, Germany, Definiens imaging GmbH 2000

[60]   Willhauck G. Comparison of object oriented classification techniques and standard image analysis for the use of change detection between SPOT multispectral satellite images and aerial photos. In Proceedings of XIX ISPRS congress. Amsterdam: IAPRS. 33: 35-42.

[61]   Laliberte AS, Rango A, Havstad KM, *et al.* Object-oriented image analysis for mapping shrub encroachment from 1937 to 2003 in southern New Mexico. Remote Sens Environ 2004; 93: 198-210.

[http://dx.doi.org/10.1016/j.rse.2004.07.011]

[62] Baatz M, Benz U, Dehghani S, Heynen M. eCognition user guide 4. Munich, Germany, Definiens imagine GmbH 2004

[63] Ryherd S, Woodcock C. Combining spectral and texture data in the segmentation of remotely sensed images. Photogramm Eng Remote Sensing 1996; 62(2): 181-94.

[64] Gregorio AD, Jansen LJM. Land Cover Classification System (LCCS): classification concepts and user manual, SDRN. Rome: FAO 1998.

[65] Kuckler AW, Zonneveld IS. Vegetation mapping. Dordrecht, the Netherlands: Kluwer Academic Publisher 1988.
[http://dx.doi.org/10.1007/978-94-009-3083-4]

[66] Olofsson P, Foody GM, Stehman SV, Woodcock CE. Making better use of accuracy data in land change studies: Estimating accuracy and area and quantifying uncertainty using stratified estimation. Remote Sens Environ 2013; 129: 122-31.
[http://dx.doi.org/10.1016/j.rse.2012.10.031]

[67] Kettig RL, Landgrebe DA. Classification of multispectral image data by extraction and classification of homogeneous objects. IEEE T Geosci Elect 1976; 1: 19-26.
[http://dx.doi.org/10.1109/TGE.1976.294460]

[68] Stuckens J, Coppin R, Bauer ME. Integrating contextual information with per-pixel classification for improved land cover classification. Remote Sens Environ 2000; 71: 282-96.
[http://dx.doi.org/10.1016/S0034-4257(99)00083-8]

[69] Sepehry A, Liu GJ. Flood induced land cover change detection using multitemporal ETM+ Imagery. Proceedings of the Center for Remote Sensing of Land Surfaces. Sept. 28-30; Bonn, Germany. 2006; pp. 1-7.

[70] Skole D, Tucker C. Tropical deforestation and habitat fragmentation in the Amazon: Satellite data from 1978 to 1988. Science 1993; 260(5116): 1905-10.
[http://dx.doi.org/10.1126/science.260.5116.1905] [PMID: 17836720]

[71] Muchoney DM, Haack BN. Change detection for monitoring forest defoliation. Photogramm Eng Remote Sensing 1994; 60(10): 1243-51.

[72] Leh M, Bajwa S, Chaubey I. Impact of land use change on erosion risk: An integrated remote sensing, geographic information system and modeling methodology. Land Degrad Dev 2013; 24(5): 409-21.
[http://dx.doi.org/10.1002/ldr.1137]

[73] Köhl M, Magnussen SM, Marchetti M. Sampling methods, remote sensing and GIS multiresources forest inventory. Berlin Heidelberg, Germany: Springer-Verlag 2006.
[http://dx.doi.org/10.1007/978-3-540-32572-7]

[74] Do Valle Júnior RF, Siqueira HE, Valera CA, *et al.* Diagnosis of degraded pastures using an improved NDVI-based remote sensing approach: An application to the Environmental Protection Area of Uberaba River Basin (Minas Gerais, Brazil). Remote Sens Appl Soc Environ 2019; 14: 20-33.
[http://dx.doi.org/10.1016/j.rsase.2019.02.001]

[75] Zhan Y, Muhammad S, Hao P, Niu Z. The effect of EVI time series density on crop classification accuracy. Optik (Stuttg) 2018; 157: 1065-72.
[http://dx.doi.org/10.1016/j.ijleo.2017.11.157]

[76] Ren H, Zhou G, Zhang F. Using negative soil adjustment factor in soil-adjusted vegetation index (SAVI) for aboveground living biomass estimation in arid grasslands. Remote Sens Environ 2018; 209: 439-45.
[http://dx.doi.org/10.1016/j.rse.2018.02.068]

[77] Loozen Y, Rebel KT, Karssenberg D, *et al.* Remote sensing of canopy nitrogen at regional scale in Mediterranean forests using the spaceborne MERIS Terrestrial Chlorophyll Index. Biogeosciences

2018; 15(9): 2723-42.
[http://dx.doi.org/10.5194/bg-15-2723-2018]

[78]   Egea-Cobrero V, Rodriguez-Galiano V, Sánchez-Rodríguez E, García-Pérez MA. Wheat yield prediction in Andalucía using MERIS Terrestrial Chlorophyll Index (MTCI) time series. Rev de Teledetección 2018; 51: 99-112.
[http://dx.doi.org/10.4995/raet.2018.8891]

[79]   Jorge J, Vallbé M, Soler JA. Detection of irrigation inhomogeneities in an olive grove using the NDRE vegetation index obtained from UAV images. Eur J Remote Sens 2019; 52(1): 169-77.
[http://dx.doi.org/10.1080/22797254.2019.1572459]

[80]   Longley PA, Goodchild MF, Maguire DJ, Rhind DW. Geographic Information Systems and Science. Chichester, UK: John Wiley 2001.

[81]   Boyd DS, Danson FM. Satellite remote sensing of forest resources: three decades of research development progress. Phys Geogr 2005; 29(1): 1-26.
[http://dx.doi.org/10.1191/0309133305pp432ra]

[82]   Peterson DL, Running SW. Applications in forest science and management. In: Asrar G, Ed. Theory and applications of optical remote sensing. New York: Wiley 1989; pp. 429-73.

[83]   Curran PJ. Imaging spectrometry. Prog Phys Geogr 1994; 18: 247-66.
[http://dx.doi.org/10.1177/030913339401800204]

[84]   Shao G, Reynolds KM. Computer applications in sustainable forest management: Including perspectives on collaboration and integration. Netherlands: Springer 2006.
[http://dx.doi.org/10.1007/978-1-4020-4387-1]

[85]   Pernar R, Storga D. Possibilities of applying GIS in forest ecosystem management. Ekologia (Bratisl) 2005; 24: 66-79.

[86]   Guyot G. Optical properties of vegetation canopies.Application of remote sensing in agriculture. London, UK: Butterworths 1990.
[http://dx.doi.org/10.1016/B978-0-408-04767-8.50007-4]

[87]   Howard JA. Remote sensing of forest resources, theory and application. London: Chapman and Hall 1991.http://www.fao.org/forestry/sofo/

[88]   Jaakkola SP. Applicability of SPOT for forest management. Adv Space Res 1989; 9(1): 135-41.
[http://dx.doi.org/10.1016/0273-1177(89)90478-X]

[89]   Strahler AH, Woodcock CE, Smith JA. Data acquisition and object modeling for industrial as-built documentation and architectural applications on the nature of models in remote sensing. Remote Sens Environ 1986; 20: 121-39.
[http://dx.doi.org/10.1016/0034-4257(86)90018-0]

[90]   Franklin SE, Wulder M, Skakun R, Carroll A. Mountain pine beetle red attack damage classification using stratified Landsat TM data in British Columbia, British Columbia, Canada. Photogramm Eng Remote Sensing 2003; 69(3): 283-8.
[http://dx.doi.org/10.14358/PERS.69.3.283]

[91]   West PW. Tree and forest measurement. 2nd ed., Berlin, Heidelberg: Springer-Verlag 2009.
[http://dx.doi.org/10.1007/978-3-540-95966-3]

[92]   Remote Sensing and Proximal Sensing (accessed on 20 May 2020) https://sphereoptics.de/ en/ application/remote-sensing-proximal-sensing/

[93]   Green Seeker System (accessed on 20 May 2020) https://agriculture.trimble.com/ product/ greenseeker-system/

[94]   Perry EM, Fitzgerald GJ, Nuttall JG, O'Leary GJ, Schulthess U, Whitlock A. Rapid estimation of canopy nitrogen of cereal crops at paddock scale using a Canopy Chlorophyll Content Index. Field Crops Res 2012; 134: 158-64.

[http://dx.doi.org/10.1016/j.fcr.2012.06.003]

[95]   Fournier RA, Luther JE, Guindon L, *et al.* Mapping aboveground tree biomass at the stand level from inventory information: test cases in Newfoundland and Quebec. Can J Res 2003; 33(10): 1846-63.
       [http://dx.doi.org/10.1139/x03-099]

[96]   Viña A, Gitelson AA, Nguy-Robertson AL, Peng Y. Comparison of different vegetation indices for the remote assessment of green leaf area index of crops. Remote Sens Environ 2011; 115(12): 3468-78.
       [http://dx.doi.org/10.1016/j.rse.2011.08.010]

[97]   Song C. Optical remote sensing of forest leaf area index and biomass. Prog Phys Geogr 2013; 37(1): 98-113.
       [http://dx.doi.org/10.1177/0309133312471367]

[98]   Toth C, Jóźków G. Remote sensing platforms and sensors: A survey. ISPRS J Photogramm Remote Sens 2016; 115: 22-36.
       [http://dx.doi.org/10.1016/j.isprsjprs.2015.10.004]

[99]   Colomina I, Molina P. Unmanned aerial systems for photogrammetry and remote sensing: A review. ISPRS J photogram remote sens2014; 92: 79-97.
       [http://dx.doi.org/10.1016/j.isprsjprs.2014.02.013]

[100] Sternberg H, Kersten T, Jahn I, Kinzel R. Terrestrial 3D laser scanning data acquisition and object modelling for industrial Aas-built documentation and architectural applications. Int Arch Photogramm Remote Sens Spat Inf Sci 2004; 35(7): 942-7.

[101] Asner GP, Bustamante MMC, Townsend AR. Scale dependence of biophysical structure in deforested areas bordering the Tapajos National Forest, Central Amazon. Remote Sens Environ 2003; 87(4): 507-20.
       [http://dx.doi.org/10.1016/j.rse.2003.03.001]

[102] Vizilter YV, Gorbatsevich VS, Vorotnikov AV, Kostromov NA. Real-time face identification *via* CNN and boosted hashing forest. Comput Opt 2017; 41(2): 254-65.
       [http://dx.doi.org/10.18287/2412-6179-2017-41-2-254-265]

[103] Ayoubi S, Shahri AP, Karchegani PM, Sahrawat KL. Application of Artificial Neural Network (ANN) to predict soil organic matter using remote sensing data in two ecosystems Biomass and remote sensing of biomass. Intechopen 2011; pp. 181-96.

[104] Fralenko VP. Spectrographic texture analysis for earth remote sensing data Artificial Intelligence and Decision Making 2010; 2: 5-11.

[105] Fralenko VP. Intelligent analysis of aerospace images using high-performance computing devices Proceedings of the conference "Artificial Intelligence: Problems and Solutions". Moscow region, Patriot Park,.

[106] Ennouri K, Triki MA, Kallel A. Applications of remote sensing in pest monitoring and crop management.Bioeconomy for Sustainable Development. Singapore: Springer 2020; pp. 65-77.
       [http://dx.doi.org/10.1007/978-981-13-9431-7_5]

[107] Yuan H, Yang G, Li C, Wang Y, Liu J, Yu H, *et al.* Retrieving soybean leaf area index from unmanned aerial vehicle hyperspectral remote sensing: Analysis of RF, ANN, and SVM regression models. Remote Sens 2017; 9(4): 309.
       [http://dx.doi.org/10.3390/rs9040309]

[108] Yi Q, Huang J, Wang F, Wang X. Evaluating the performance of PC-ANN for the estimation of rice nitrogen concentration from canopy hyperspectral reflectance. Int J Remote Sens 2010; 31(4): 931-40.
       [http://dx.doi.org/10.1080/01431160902912061]

[109] Şenkal O. Modeling of solar radiation using remote sensing and artificial neural network in Turkey. Energy 2010; 35(12): 4795-801.
       [http://dx.doi.org/10.1016/j.energy.2010.09.009]

[110] Şahin M. Modelling of air temperature using remote sensing and artificial neural network in Turkey. Adv Space Res 2012; 50(7): 973-85.
[http://dx.doi.org/10.1016/j.asr.2012.06.021]

SUBJECT INDEX

A

www.ingramcontent.com/pod-product-compliance
Lightning Source LLC
Chambersburg PA
CBHW041710210326
41598CB00007B/607